SPACE SHUTTLE

SPACE
SHUTTLE

BILL YENNE

GALLERY BOOKS
An imprint of W.H. Smith Publishers Inc.
112 Madison Avenue
New York, New York 10016

A Bison Book

Published by Gallery Books
A Division of W H Smith Publishers Inc.
112 Madison Avenue
New York, New York 10016

Produced by
Bison Books Corp.
17 Sherwood Place
Greenwich, CT 06830

ISBN 0-8317-7989-6

Printed in Hong Kong

1 2 3 4 5 6 7 8 9 10

Page 1: The Space Shuttle *Columbia* is eased
onto Pad 39A at Kennedy Space Center in
preparation for one of the early STS missions.

Page 2-3: The Space Shuttle *Discovery* was
launched for the first time on 30 August 1984
and carried the first astronaut to represent
private industry, Charles Walker of McDonnell
Douglas.

Below: Like a ghost ship adrift in a sea of mist,
the Orbiter *Challenger* slowly makes its way
toward launch pad 39A at Cape Canaveral.

Overleaf: Painters at Rockwell's Palmdale plant
put the finishing touches on the flag that
adorned *Challenger's* port wing.

Contents

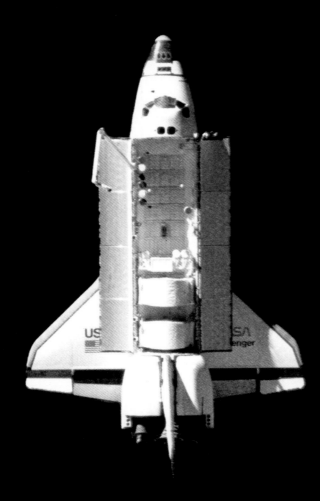

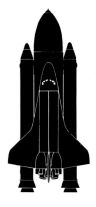

Prologue

The Space Shuttle is an entirely new breed of manned spacecraft. No other has ever been like it—a reusable space plane capable of going into space again and again, carrying a variety of payloads and numerous crew members. The Space Shuttle Transportation System (STS), or the Space Shuttle program as it is commonly known, is the most important earth-orbit space program ever undertaken by the United States.

The first manned spacecraft used by the Americans and the Russians were called ballistic re-entry capsules, archaic expendable containers that were each good for a single space flight. Between 1961, when both the US and the USSR began their manned space programs, and 1969, both countries fielded three generations of ballistic re-entry capsules. The Americans began with the one-man Mercury and progressed to the two-man Gemini in 1965. They first tested the three-man Apollo capsule in 1968, and a year later the Apollo spacecraft took the first of six American crews to a landing on the moon.

The Soviet Union sent five men and a woman into space aboard its single-crew Vostok capsules between 1961 and 1963. In 1964 and 1965 five men went into space on two missions in the Voskhod spacecraft, a

Challenger, photographed from the unmanned free-flying German SPAS-01 spacecraft in June 1983 with its payload bay doors open exposing pallets and cradles. The nonpressurized 15-by-60-foot payload bay can accommodate a wide variety of cargo, experiments and satellites that are operated either by payload specialists or the remote manipulator arm, which is seen at the top left of the bay.

modified Vostok designed for multiple crews. The third generation Soviet capsule, Soyuz, came on line in 1967, but the first flight ended in the tragic death of its single crewman. Undaunted, the Russians pressed forward with their Soyuz program.

The historic rendezvous in space between a Soviet Soyuz capsule and an American Apollo capsule took place in 1975. After the Apollo-Soyuz mission, the nineteenth in the Soyuz series, the Soviet Union continued its ballistic re-entry capsule manned space-flight program, and completed nearly four dozen manned Soyuz missions by the mid-1980s. For the United States, however, the Apollo-Soyuz project was the final flight of an Apollo spacecraft and the last space mission by an American ballistic re-entry capsule. During the next six years, no American ventured into space while NASA turned its attention to building the next generation of manned spacecraft—a plan that had been initiated by NASA and the Defense Department in 1969, the year Neil Armstrong set foot on the moon.

The idea of a reusable space plane was not new, however. In Germany during World War II, Dr Eugen Sanger tried unsuccessfully to sell such a concept to Hitler as a bomber. After the war, the complexities of com-

Astronauts Gene Cernan and Thomas Stafford prepare to squeeze out of the close confines of the Gemini 9 spacecraft on 17 May 1966, shortly after learning that the Agena target vehicle with which they were to have docked failed to achieve orbit. This was the second attempt at a docking in space. The two men finally went into space on 1 June for a successful five-day flight, the seventh for the series of two-man Gemini spacecraft.

bining aerodynamic wing surfaces with the thrust needed for space flight forced engineers, for expediency, to use rockets and eventually manned ballistic re-entry capsules. During the 1950s and 1960s NASA and the US Air Force began some tentative research into the idea of a reusable space plane. From North American Aviation they commissioned the X-15s, a series of rocket-powered aircraft capable of speeds nearly seven times the speed of sound, which flew higher than 60 miles during a series of 199 test flights between 1959 and 1963. From Boeing they commissioned the X-20, the cornerstone of the Dynamic Ascent and Soaring Flight project, which bore the unfortunate nickname Dyna-Soar. Canceled in 1962, the X-20 was to have been the first practical space plane that could be launched into space and return to earth like an airplane. With the extinction of the X-20 Dyna-Soar, attention turned to a larger and thus more practical space plane.

The final design of the Space Shuttle was approved in 1972 and construction started. The Shuttle dream became a reality on 12 April 1981, the twentieth anniversary of the first manned space mission, when the launch of the Orbiter *Columbia* and its two-man crew ushered in a new era in space flight.

A view of the Apollo 17 Command and Service Module (CSM) from the Lunar Module (LM) *Challenger* during rendezvous and docking maneuvers in lunar orbit, with its scientific instrument bay exposed. *Challenger* carried Gene Cernan and Harrison Schmidt, the last two men to set foot on the lunar surface. Two of the given names used on Apollo spacecraft were re-applied to Shuttle Orbiters. *Columbia* was the Apollo II CSM and Orbiter OV-102, while *Challenger* was also the name of Orbiter OV-99.

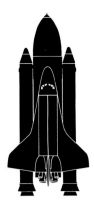

A Short History

The X-15 and X-20 projects during the 1950s and 1960s provided the technical foundation for the Space Shuttle and the impending end of the Apollo program provided the impetus to proceed. By 1972 the Apollo moon flights were over and the Space Shuttle Transportation System program had begun. Rockwell International (the successor to North American Aviation, which had built the X-15) was commissioned to design and build the space plane portion of the system. Morton Thiokol built the special launch vehicles, and Martin Marietta built the huge fuel tank, the largest single part of the Space Shuttle. The original timetable called for the first flight of the system in 1978 and for over a hundred missions to be completed by 1984. The schedule slipped badly, however, and the first space plane, or Orbiter, wasn't even ready for initial testing until 1977.

This first Orbiter, designated OV-101 (Orbiting Vehicle, first), was named *Enterprise* after the mythical Hollywood starship of the television series *Star Trek.* Designed solely for testing the Orbiting Vehicle's aerodynamic flight properties within the earth's atmosphere, *Enterprise* would never reach the airless world of space, only the thin air over California's high desert. *Enterprise* was trucked out to Edwards AFB from Rockwell's Palm-

(continued on page 17)

A new generation of Americans grew up with a new generation of manned spacecraft. In this mother's lifetime, Americans had made their first tentative steps into space and had gone to the moon. During her young daughter's lifetime, this strange airplane will revolutionize the capability of Americans to live and work in space. Perhaps one day she, too, will venture beyond the bonds of earthly gravity.

The first steps of each Orbiter's million-mile odyssey begin behind a trailer truck on the back roads of Los Angeles County. The Orbiter *Enterprise* is shown here en route from Palmdale to Edwards AFB in 1977, accompanied by a caravan of well wishers.

dale, California plant in early 1977, where it was hoisted aboard a big former American Airlines 747 upon whose back it would make its first flight. For the first tests, the Orbiter remained attached to the 747, but finally, on 12 August 1977, it was released and glided free for the first time. *Enterprise* made five successful glide tests in California and in 1979 was brought to Kennedy Space Center at Cape Canaveral for further testing.

Meanwhile in 1979, *Columbia* (OV-102) was completed and also delivered to NASA's Kennedy Space Center. The first Shuttle Orbiter built intentionally for space flight, *Columbia* first flew in April 1981 and again in November. In 1982 *Columbia* completed a four-flight test series and flew its first operational mission, which entailed launching a pair of communications satellites.

Although the Space Shuttle Transportation System was originally conceived as an adjunct to the manned space-station project, the Shuttle Orbiters successfully turned to duty launching satellites and providing a platform for zero-gravity scientific experiments when NASA's space station failed to materialize. *Columbia*'s November 1983 flight, its last for over two years, marked the longest ever Shuttle mission (10 days) as well as the first flight of a non-American astronaut, Ulf Merbold, aboard an

(continued on page 29)

The Orbiter *Enterprise* during assembly at Rockwell's Palmdale, California plant. *Enterprise*, designated OV-101, was the only Orbiter to have been built with a nose probe. *Overleaf:* Enterprise is mated to the 747 for its maiden glide through the thin, high desert air. The big white tail cone was used to cover the engines for the early glide tests. The cone was removed after the early tests and the nose probe was deleted after the flight-test phase.

Enterprise touches down at Edwards AFB after one of its last glide tests. The spacecraft was never intended for actual space flight, but for tests that would be conducted in earth atmosphere. *Overleaf: Columbia* waits on Pad 39A at Kennedy Space Center in April 1981, prior to its first launch.

Columbia on the pad (*above*) and *Columbia*'s first launch (*right*), on 12 April 1981. On board mission STS-1 was a two-man crew, Robert Crippen and John Young, typical of the early missions. The white external fuel tanks were used only for the first two Shuttle flights; thereafter the tanks were painted only with rust-colored primer, saving half a ton in weight.

Columbia on the ground at Edwards AFB after one of its early flights. A freak cloudburst, unusual but not unheard-of in the high desert, had turned part of the usually dry lakebed into a lake.

American spacecraft. Merbold, a German, went along to help operate the European Space Agency's Spacelab module, a zero-gravity scientific laboratory designed to fit the Orbiter's payload bay.

April of 1983 saw the first flight of *Challenger,* and during that year the new Orbiter was used for three out of the four Shuttle missions that were flown. Designated OV-99, *Challenger,* like *Enterprise,* had not originally been intended for actual space flight, but it was later modified. Ironically, *Challenger* quickly became NASA's bird of choice, making more space flights in its three long years of service than any of its sister ships. It was ironic, too, that *Challenger* was the first Orbiter to be lost in service.

There were five Shuttle flights in 1984. In that year the third space-rated Orbiter, *Discovery* (OV-103), made its debut in August and its second flight in November. Of the 10 operational Shuttle flights that had taken place between November 1982 and the end of 1984, *Challenger* had made six and *Columbia* and *Discovery* two each. During the 26 months since *Columbia*'s first operational flight, the Space Shuttle program had put 49 astronauts into space, more than had flown during the entire Apollo program. These people included the first, second, third and fourth American women to fly in space. Sally Ride was the first American woman to

(continued on page 39)

The Orbiter *Challenger* goes into space on one of its early flights. Sally Ride (*overleaf*) rode aboard *Challenger*'s second flight, STS-7, to become the first American woman in space. She is shown floating freely on the flight deck, performing her mission specialist duties — communicating with ground controllers in Houston and handling reference data, hand calculators and other aids all at once.

Above: Astronaut James van Hoften in the cargo bay of the Space Shuttle *Challenger* in April 1984. During mission 41-C, van Hoften and George Nelson were outside the spacecraft for 7 hours 18 minutes, the longest extravehicular activity since Apollo 17. Using their manned maneuvering units, they retrieved and repaired the Solar Maximum Mission ('Solar Max') Satellite seen here behind van Hoften, in the rear of the cargo bay. *Right:* Astronaut Bruce McCandless II moves in to conduct a test on the Shuttle pallet satellite during 41-B extravehicular rehearsal for the repairs that were to be made to Solar Max during the next Shuttle mission.

fly in space (June 1983) and the first woman to fly in space a second time (October 1984). Pilot Robert Crippen flew into space aboard *Challenger* three times during that period, while John Young, who had logged two missions each with Gemini and Apollo, flew two more aboard the Shuttle and became the first person to fly in space six times.

The year 1985 saw a record nine flights by Shuttle Orbiters including four by *Discovery,* three by *Challenger* and two by the fourth and last Orbiter, *Atlantis* (OV-104), which made its maiden voyage in October. *Columbia* was supposed to have made a flight in December, giving the program 10 flights for the calendar year, but weather delays put that mission off until 12 January 1986. Had *Columbia* been launched as planned, 1985 would have seen all four Orbiters in space during the same year. Unknown at year's end, that possibility would soon be lost forever.

After the successes of 1985, the year 1986 promised to be the best yet for the Shuttle program. There were 15 missions scheduled in addition to the one that had slipped over from December 1985. Included were observations from space of Halley's Comet during its once-in-76-years turn past the sun, the launch of the Galileo spacecraft that would penetrate Jupiter's atmosphere and the launch of the Hubble Space Telescope.

(continued on page 49)

Left: The Orbiter *Challenger* in earth-orbit with its payload bay doors open during 41-B extravehicular activity. Bruce McCandless took this photograph from a camera fixed on his helmet as he traveled freely in space using the manned maneuvering unit. Robert Stewart is shown beneath the remote manipulator arm.

Overleaf: F Story Musgrave, STS-6 mission specialist, translates down the earth-orbiting *Challenger's* payload bay door hinge line with a bag of latch tools. His task here was to evaluate the techniques required to move along the payload bay's edge with tools.

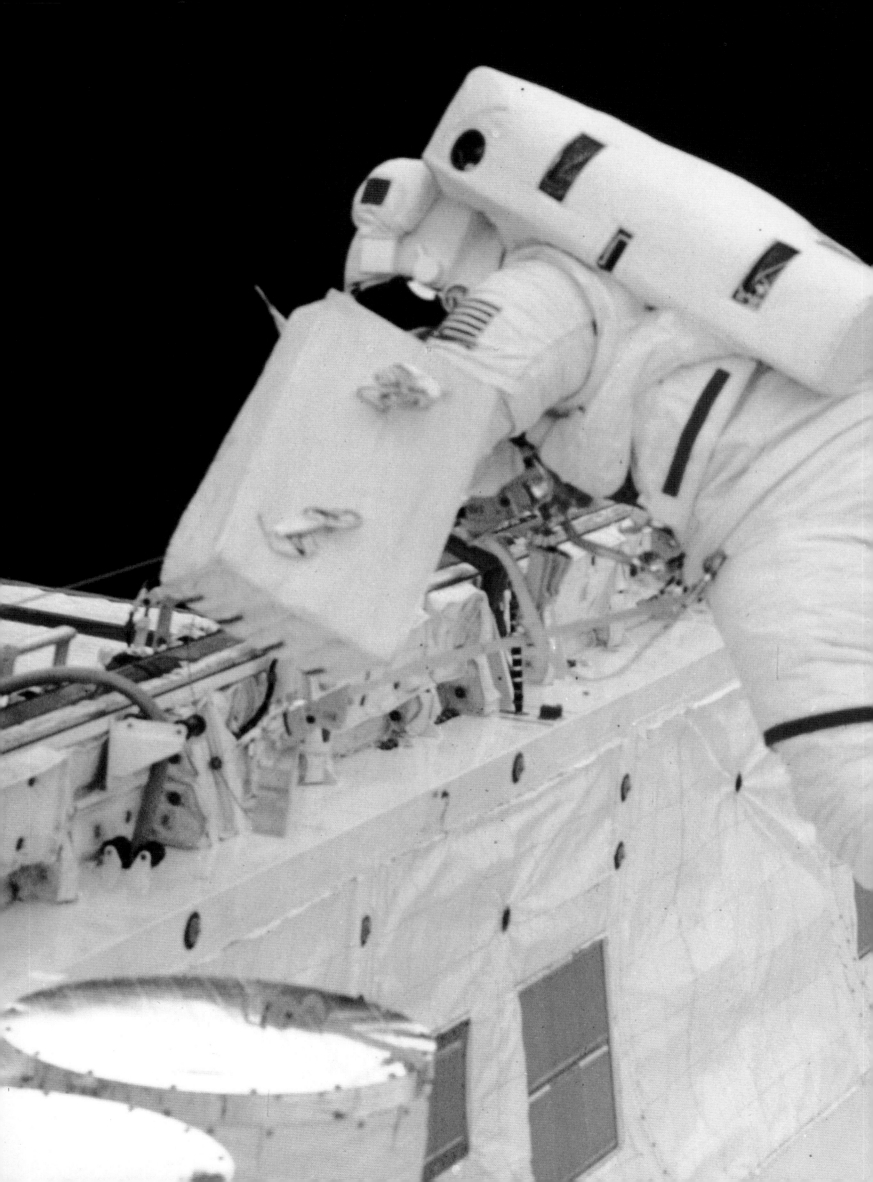

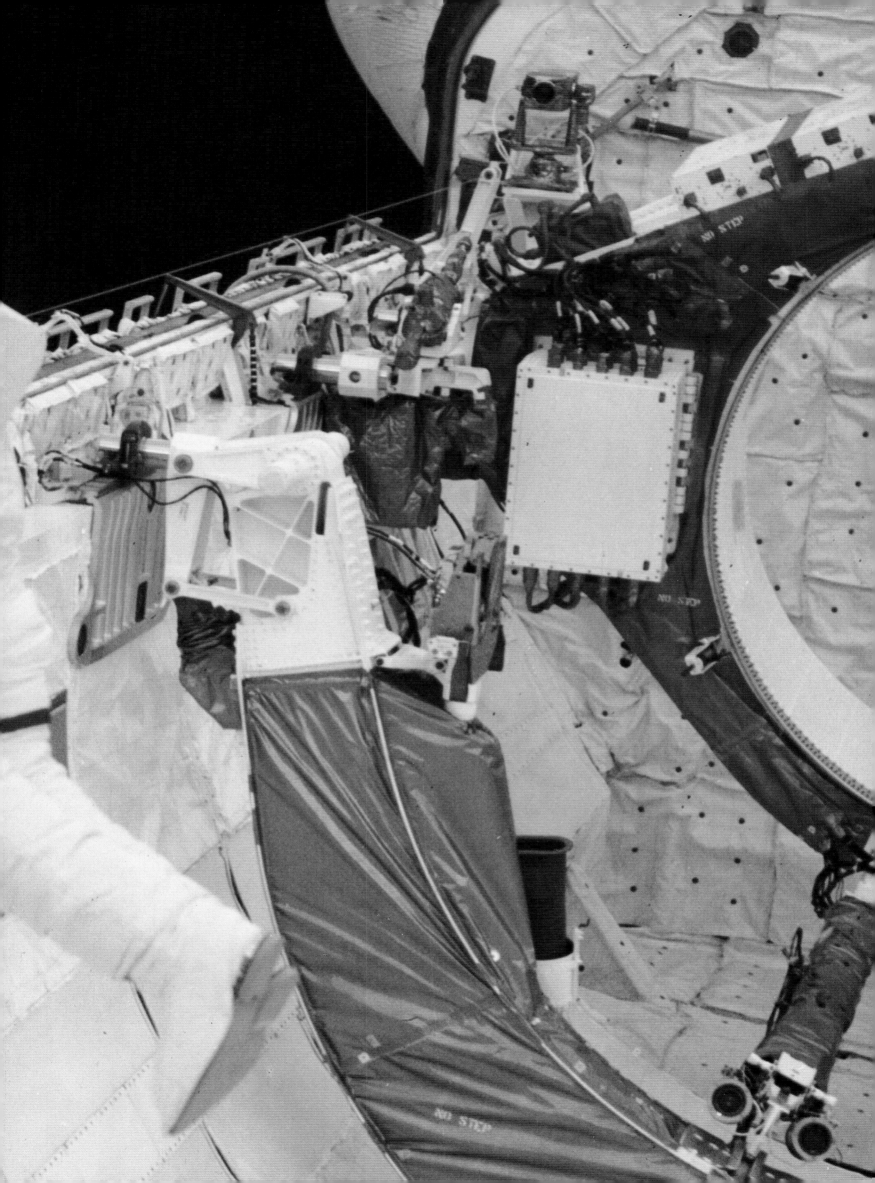

With seven crew members and a variety of international payloads on board, Space Shuttle *Discovery*, attached to two solid rocket boosters and an external fuel tank, pushes away from the launch pad on 17 June 1985 to begin mission 51-G.

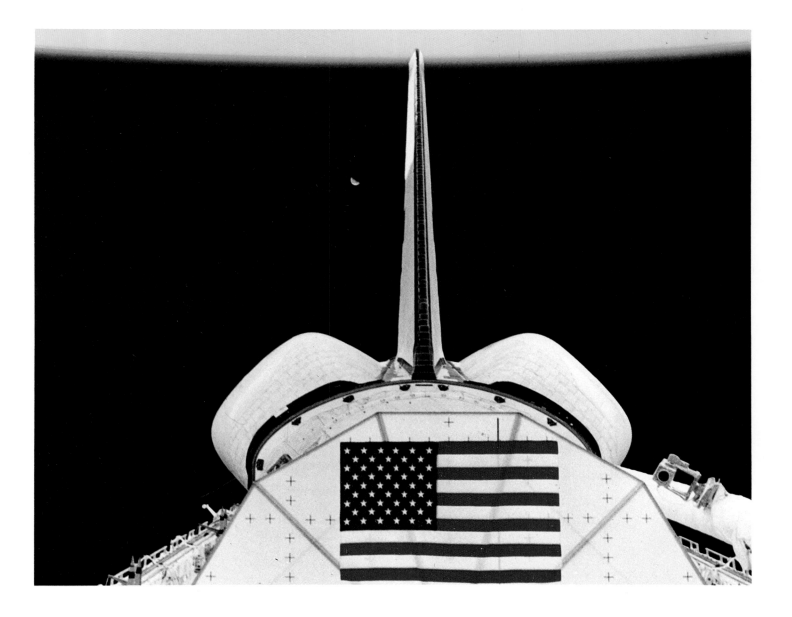

Left: Mission 41-B astronaut Bruce Mc-Candless leans out into space with his feet anchored in the mobile foot restraint connected to the remote manipulator system, called by some NASA's first 'cherry-picker' in space. *Above:* An astronaut's-eye-view of two thirds of the cargo bay, the white and blue horizon scene of the earth and a gibbous moon photographed from the aft flight deck of *Challenger* during STS-8.

Challenger touches down for the last time on 6 November 1985. The weeklong 61-A/Spacelab D-1 mission carried a record crew of eight — Henry Hartsfield, Jr; Steven Nagel; Bonnie Dunbar; James Buchli; Guion Bluford, Jr, all of the United States; Reinhard Furrer and Ernst Messerschmid of West Germany and Wubbo Ockels of the Netherlands. This was the first space flight that involved the citizens of three nations.

At the start of 1986, NASA for the first time had a full fleet of four Orbiters ready to meet the challenge of a 15-mission year. For two years, from November 1983 to October 1985, *Challenger* and *Discovery* had carried the entire load of the Shuttle program. *Atlantis* came on line in October and two months later *Columbia* was back in service.

The first flight of this busiest of seasons was scheduled for 26 January 1986, just a week after *Columbia*'s return. The mission was particularly auspicious because it was to be the first to carry a private citizen who was neither a politician nor an aerospace engineer. That person was Christa McAuliffe, a 37-year-old high school social studies teacher who would teach the first classroom lesson to be broadcast from outer space. The Orbiter was *Challenger*, NASA's favorite, the Orbiter of choice. The first flight of the 1986 schedule was intended to be the first in a new era of citizen participation in space travel, but instead it proved to be the end of an era.

After several weather-related delays, *Challenger* lifted off the Cape Canaveral launch pad for the tenth and final time at 11:38 am on 28 January 1986. Christa McAuliffe's students in New Hampshire and students across the nation watched as the crisp white bird rose into the blue sky

(continued on page 53)

Sharon Christa McAuliffe, mission 51-L citizen payload specialist representing the Teacher in Space Project, poses near launch Complex 39 prior to the launch of 61-B in November 1985 at Kennedy Space Center.

Overleaf: Flight directors Jay Greene (*right*) and Alan Briscoe study data on monitors at their consoles in the flight control room at Johnson Space Center's mission control center. The photograph was made just moments after the announcement came that *Challenger*'s launch phase was 'not nominal.' As far as they had seen, the mission 51-L launch had gone flawlessly, but 73.63 seconds after that launch all the telemetry from *Challenger* suddenly stopped.

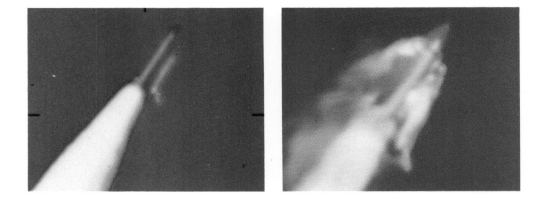

clutching its fuel tank. It was a beautiful day and a perfect launch, the Shuttle program's twenty-fifth. Unbeknownst to the crew, the ground controllers or observers, something was terribly, horribly wrong.

Flames licked out of one of the huge solid rocket boosters that propelled *Challenger* upward. Swiftly they grew into a sheet of flames and 73 seconds after launch, the liquid hydrogen and liquid oxygen in the fuel tank ignited in a gargantuan fireball that destroyed both the tank and *Challenger* instantly. The two solid rocket boosters twisted across the sky like blinded, frightened animals.

The mission that had been earmarked to confirm that space travel had become safe and routine served to remind us that space as a frontier was still untamed. Seven lives were lost, but other brave men and women stood ready to accept the challenge and see the Shuttle program forward into its second era.

(continued on page 57)

The fiery end of *Challenger* and 51-L: At 60.6 seconds, a small tongue of flame was visible at the base of the right solid rocket booster as an O-ring seal failed (*above left*). Two seconds later, the booster slammed the side of the external fuel tank, rupturing it and causing a leak of liquid hydrogen and liquid oxygen. Ten seconds later, the leak from the fuel tank was a torrent and at the 73.2-second mark (*above right*) it ignited and began to consume *Challenger* and her crew. At the 73.6-second mark, a huge explosion occurred, ripping the Orbiter into thousands of fragments while the two solid rocket boosters careened across the sky like blinded, panicked animals, only to be destroyed 37 seconds later by ground controllers when the boosters began to arc back toward populated areas. The photograph at the left clearly shows the path of the Shuttle before the explosion, the explosion itself and the paths of the two boosters afterward.

Overleaf: Enterprise on the pad at Vandenberg AFB prior to its retirement in 1985. It was used to help prepare the Vandenberg launch facility. Had it not been for the 51-L disaster, Vandenberg AFB would have seen its first Shuttle launch in June 1986.

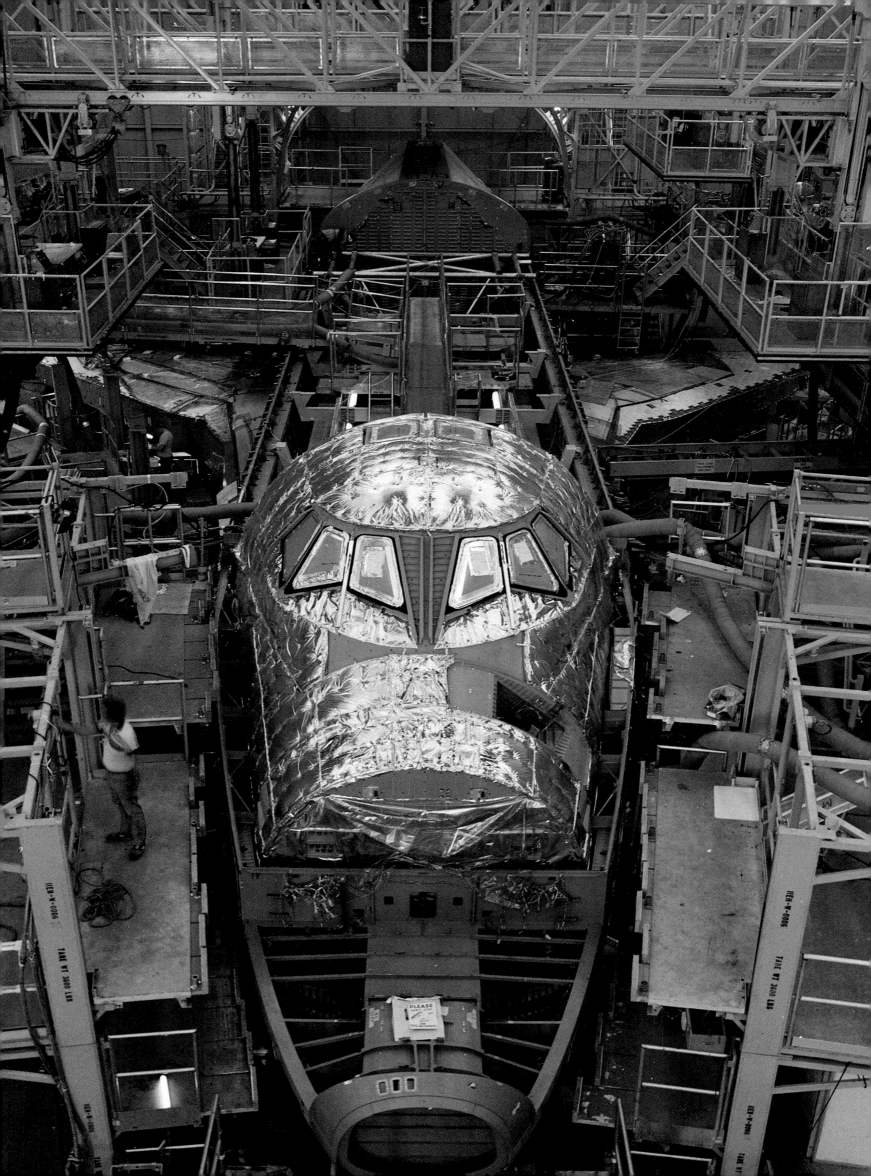

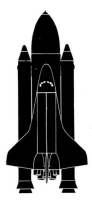

Basic Components

Each Space Shuttle Transportation System spacecraft is composed of four parts: an external fuel tank, two solid rocket boosters and an Orbiting Vehicle, or Orbiter. The overall spacecraft stands 184 feet tall and weighs more than four million pounds fully loaded and fully fueled.

The largest part of the system is the external fuel tank, which is 154 feet long and 27½ feet in diameter and which holds more than a half-million pounds of liquid hydrogen fuel and liquid oxygen oxydizer. Built by Martin Marietta, the fuel tanks are the only part of the system not designed for reuse. The two Morton Thiokol solid rocket boosters are the first such launch vehicles ever designed for reuse. With a total thrust of over five million pounds, the Shuttle's solid rockets constitute the second most powerful launch vehicle ever built, exceeded in power only by the Saturn 5 vehicle used to send the Apollo astronauts to the moon.

The centerpiece of the whole system, of course, is the Orbiter, the 122-foot-long space plane that carries the crew and payload into space as a spacecraft and returns to earth as an airplane. Buoyed aerodynamically on wings with a 78-foot span, the Orbiter glides to earth for an airplane-type landing. It is the first, and to date only, spacecraft that has

(continued on page 65)

Challenger under construction at Rockwell International's Palmdale plant: The forward fuselage skeleton after the installation of windows and before completion of the bulkheads. The Shuttle Main Engines are still to come and the payload bay is just a shell.

Left: The upper forward fuselage is balanced on top of *Challenger*'s crew cabin. Five years later this crew cabin became the tomb of the seven brave souls of mission 51-L. The above cutaway view of the Orbiter shows flight deck detail as well as structural details of the fuselage, wing and tail. The narrow blue structure running parallel to the payload bay is the remote manipulator arm in stowed position. The blue tanks are the reaction control system fuel tanks and the yellow ones are for the Orbital Maneuvering System.

Overleaf: Rockwell workers inspect the tiles aft of *Challenger*'s port wing root. This may look like tiling your bathroom, but while the process is similar, each tile here is precision made and probably more expensive than your entire bathroom.

Page 62-63: Challenger's starboard mid-fuselage with most of the tiles in place. Among those yet to be installed are the ones that surround the payload bay hinges.

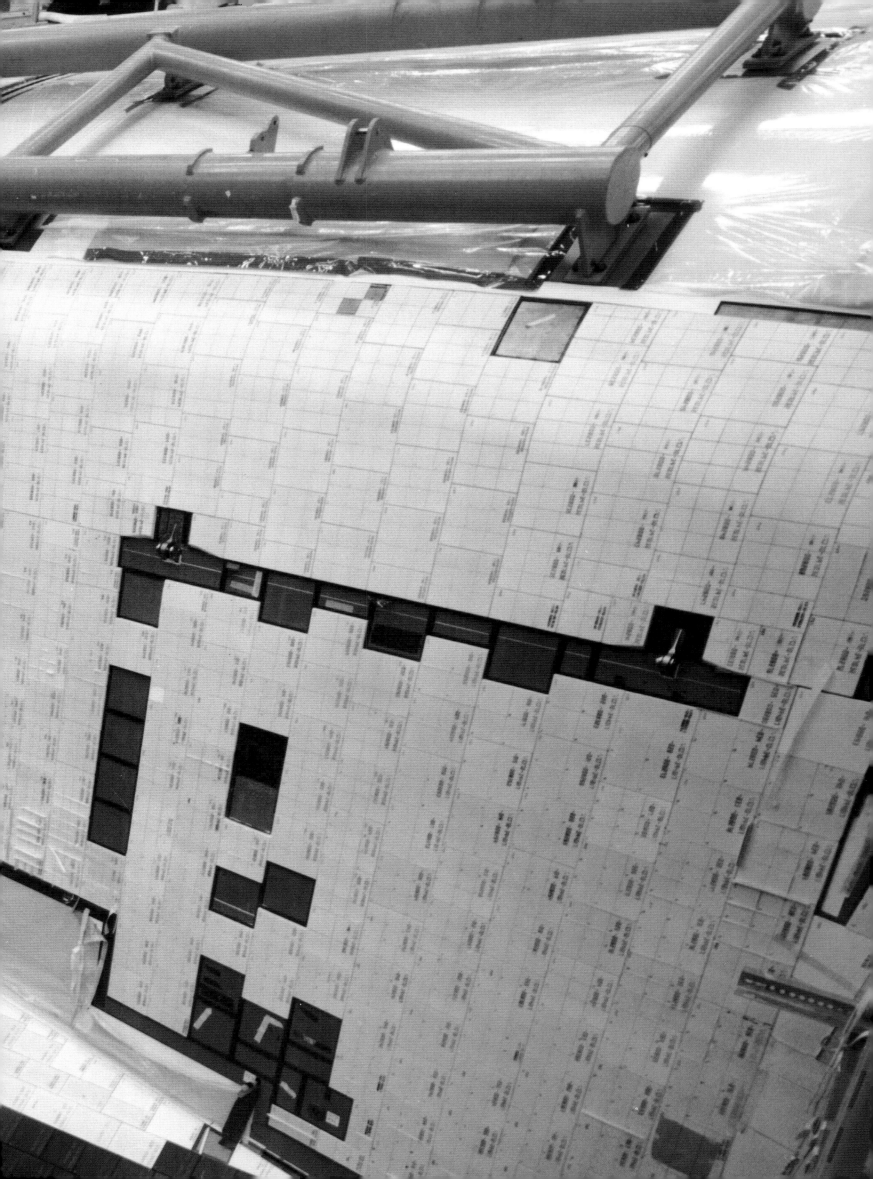

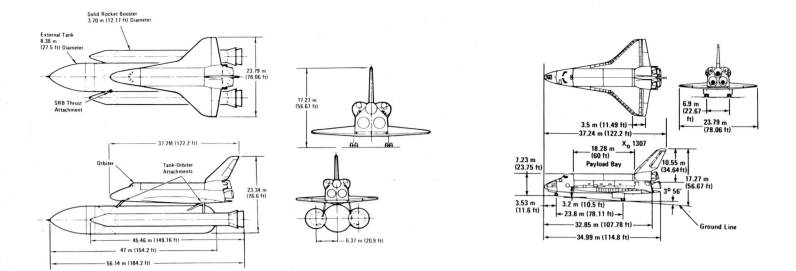

been capable of such a controlled landing. By contrast, the ballistic re-entry capsules used for other manned missions returned to earth dangling passively beneath parachutes.

The Orbiters had a unit cost of roughly $1.2 billion, although at the time *Challenger* was lost in early 1986, the estimated cost of reopening the assembly line to build a replacement (OV-105) was estimated at more than $1.5 billion. Despite its high initial cost, the Shuttle Orbiters over the long term can launch satellites much more cheaply than expendable rockets.

Roughly the same size as a commercial jetliner such as a DC-9 or a 727, the Orbiter is constructed of an aluminum framework. Its exterior is covered with thermal protective materials to shield it from solar radiation and the extreme heat of atmospheric re-entry. The top of the Orbiter is covered with coated silica and the sides with flexible sheets that can protect them at temperatures up to 1200 and 700 degrees Fahrenheit, respectively. The bottom of the Orbiter and the leading edge of the tail (which take most of the friction of re-entry) are covered with glossy black silica tiles that protect them at temperatures up to 2300 degrees Fahrenheit. A black reinforced carbon material provides the same level of protection for the Orbiter's nose and the leading edge of the wing.

(continued on page 73)

The Orbiter *Enterprise* is attached to the fuel tank and the solid rocket boosters. The Orbiter-External Tank Separation System separates the Orbiter from the external tank at three structural points and two umbilicals. Separation of the Orbiter and external tank is triggered automatically by Orbiter computers. The drawings (*above*) show the dimensions of the entire Space Shuttle system (*left*) and of the Orbiter (*right*).

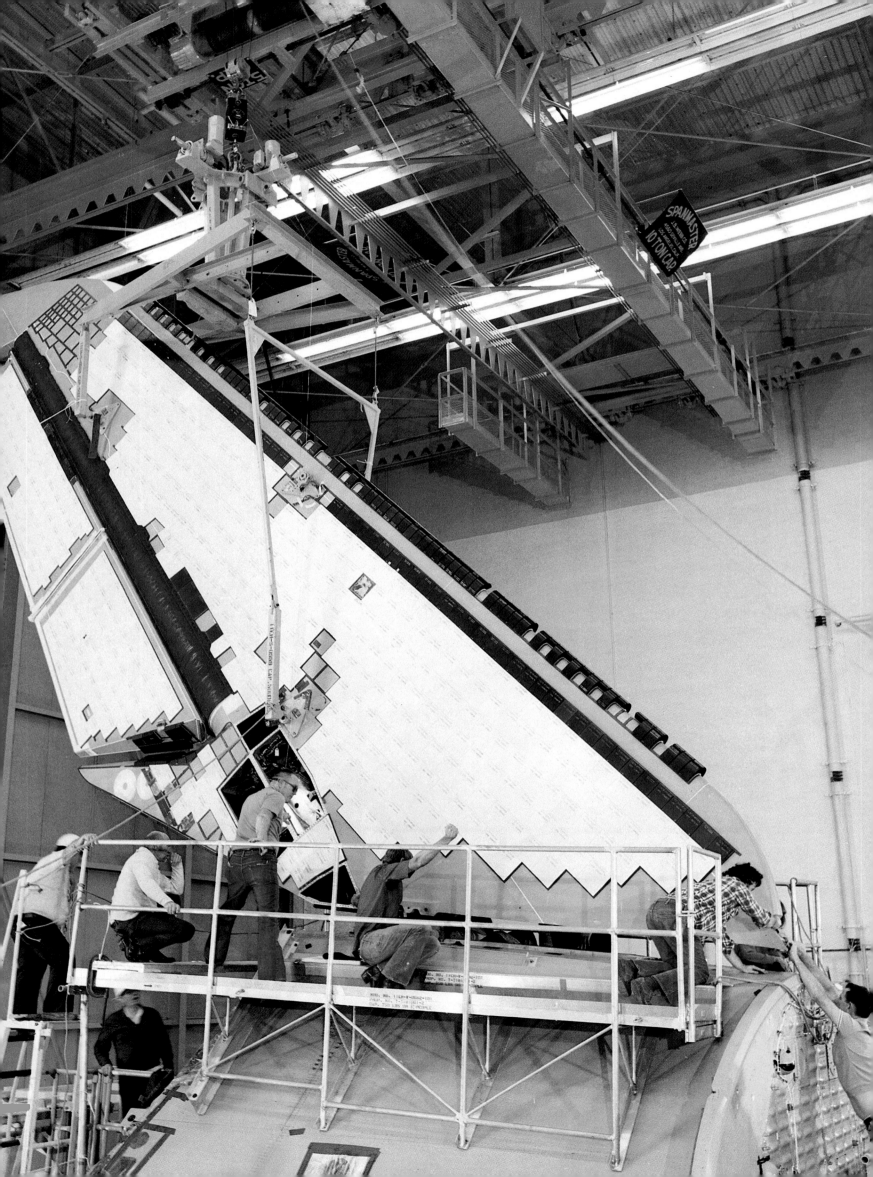

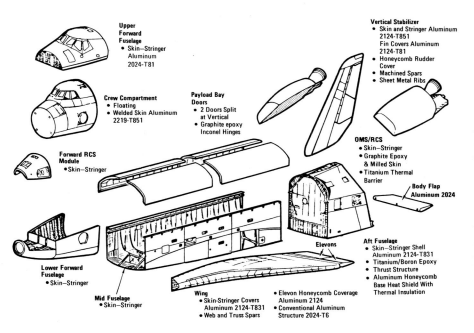

Upper Forward Fuselage
- Skin—Stringer Aluminum 2024-T81

Crew Compartment
- Floating
- Welded Skin Aluminum 2219-T851

Forward RCS Module
- Skin—Stringer

Lower Forward Fuselage
- Skin—Stringer

Mid Fuselage
- Skin—Stringer

Payload Bay Doors
- 2 Doors Split at Vertical
- Graphite epoxy Inconel Hinges

Vertical Stabilizer
- Skin and Stringer Aluminum 2124-T851 Fin Covers Aluminum 2124-T81
- Honeycomb Rudder Cover
- Machined Spars
- Sheet Metal Ribs

OMS/RCS
- Skin—Stringer
- Graphite Epoxy & Milled Skin
- Titanium Thermal Barrier

Body Flap
Aluminum 2024

Elevons

Aft Fuselage
- Skin—Stringer Shell Aluminum 2124-T831
- Titanium/Boron Epoxy
- Thrust Structure
- Aluminum Honeycomb Base Heat Shield With Thermal Insulation

Wing
- Skin-Stringer Covers Aluminum 2124-T831
- Web and Truss Spars
- Elevon Honeycomb Coverage Aluminum 2124
- Conventional Aluminum Structure 2024-T6

Far left: Rockwell workmen assembling the vertical stabilizer for *Columbia.* The vertical tail consists of a structural fin surface, the rudder/speed brake surface, a tip and a lower trailing edge. The rudder splits into two halves to serve as a speed brake. *Above:* The upper forward fuselage for *Challenger* is lifted onto the crew compartment. The drawings *(left)* show details of the construction of the Orbiter and the materials used. *Overleaf:* A closeup view of *Challenger*'s payload bay taken during mission STS-7 from the temporarily free-flying Shuttle pallet satellite.

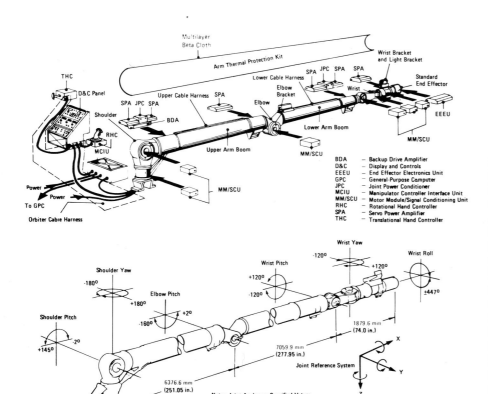

BDA	— Backup Drive Amplifier
D&C	— Display and Controls
EEEU	— End Effector Electronics Unit
GPC	— General-Purpose Computer
JPC	— Joint Power Conditioner
MCIU	— Manipulator Controller Interface Unit
MM/SCU	— Motor Module/Signal Conditioning Unit
RHC	— Rotational Hand Controller
SPA	— Servo Power Amplifier
THC	— Translational Hand Controller

Far left: Having just assisted in the successful capture of the Westar 6/PAM-D spacecraft, secured here in *Discovery*'s cargo bay, astronaut Joseph Allen rides a cherry picker over to a stowage area/work station to wrap up extravehicular activities during mission 51-A. *Above:* The handlike, end effector of the remote manipulating system shown here was at the command of astronaut Anna Fisher during much of the eight-day 51-A mission. She manipulated the arm during the capture of two stranded communications satellites by Allen and Dale Gardner. The components of the remote manipulator system are shown in the drawing at the left.

71

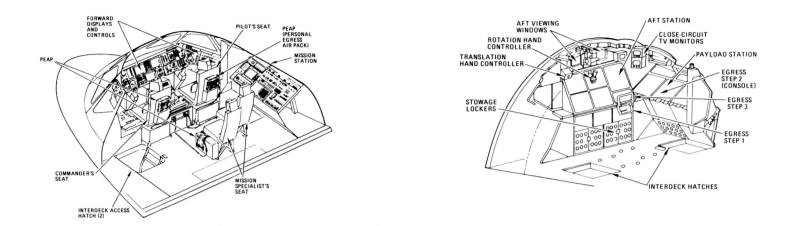

On both *Discovery* and *Atlantis* many of the tiles have been replaced by fibrous refractory composite insulation or silica fiber blankets. This resulted in an unloaded, unfueled weight of 147,980 pounds compared to 152,128 pounds for *Columbia*, the first Orbiter to fly. The saving in weight has permitted the newer Orbiters to carry roughly a ton of additional payload weight. The Orbiter payload, which can exceed 32 tons, is carried in the Orbiter's payload bay, which measures 60 by 15 feet. The payload bay can accommodate up to five satellites on pallets, the European Space Agency Spacelab module or a mix of cargo.

In the forward part of the Orbiter, ahead of the payload bay, is the flight deck and crew cabin. It is in this area that the crew lives and works while the spacecraft is in orbit. It is pressurized to 14.7 pounds per square inch to simulate sea level rather than the 5 pounds per square inch used on earlier US spacecraft. The atmosphere within the cabin is an oxygen-nitrogen mix that also simulates the earth's atmosphere rather than the pure oxygen atmosphere of earlier American manned spacecraft.

The Orbiter flight crew consists of a pilot and a spacecraft commander, either of whom can fly the ship, which can be flown and landed by one person if necessary. Only these two persons are required for the basic

(continued on page 81)

Left: The forward display controls in operational mode. The top forward crew station (*above left*) is for the pilot and commander, and the aft crew station (*above right*) is where the mission specialists work. *Overleaf:* The cockpit of *Enterprise* under construction, prior to installation of the three video displays, seats and starboard control stick.

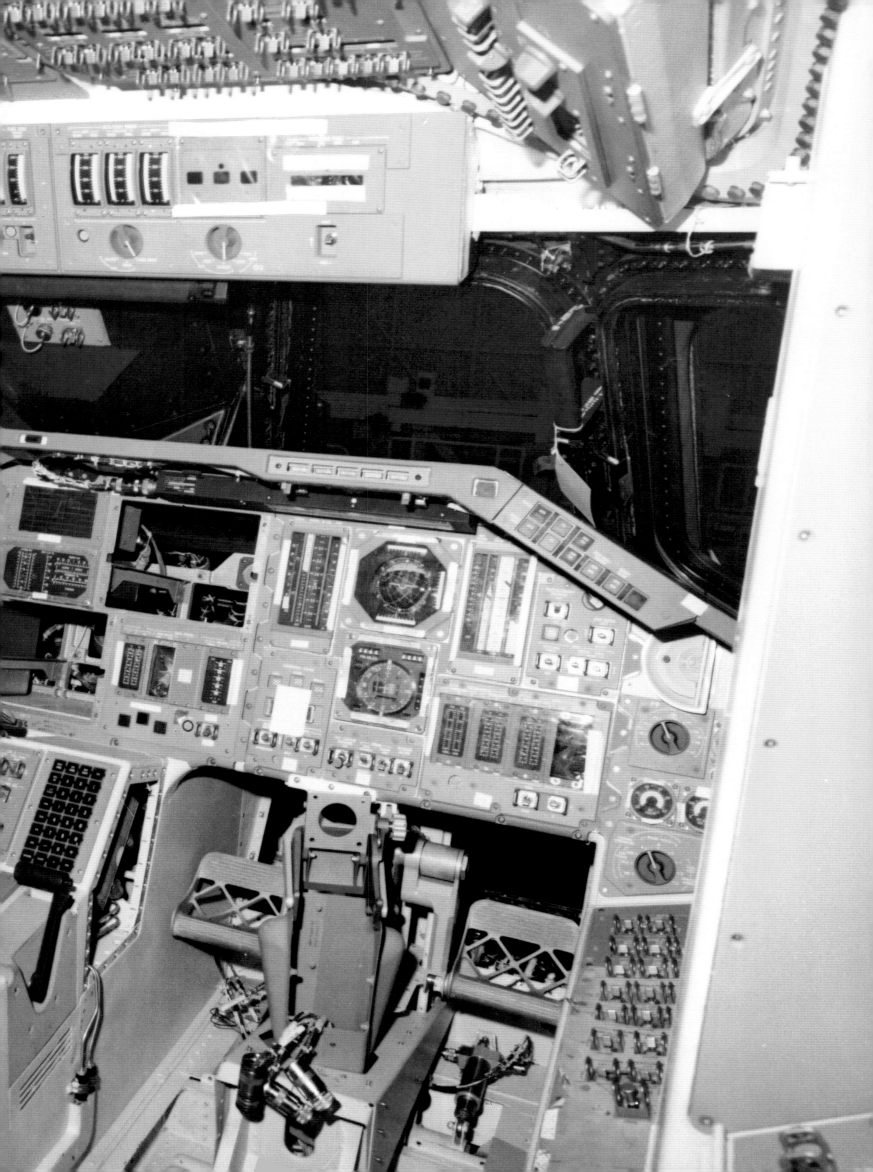

At work in the aft station. Robert Overmyer (*left*), pilot for STS-5 in 1982, behind the pilot's seat on the flight deck of the Space Shuttle *Columbia*. This position is the one used by the four crew members for observing and photographing the earth through the ceiling windows. *Above:* Astronaut Joseph Allen, STS-5 mission specialist, lets a spot meter float free for a moment during a period devoted to out-the-window photographs of earth. He is on the flight deck, behind the pilot's station. *Overleaf:* The aft crew station of *Enterprise* during assembly.

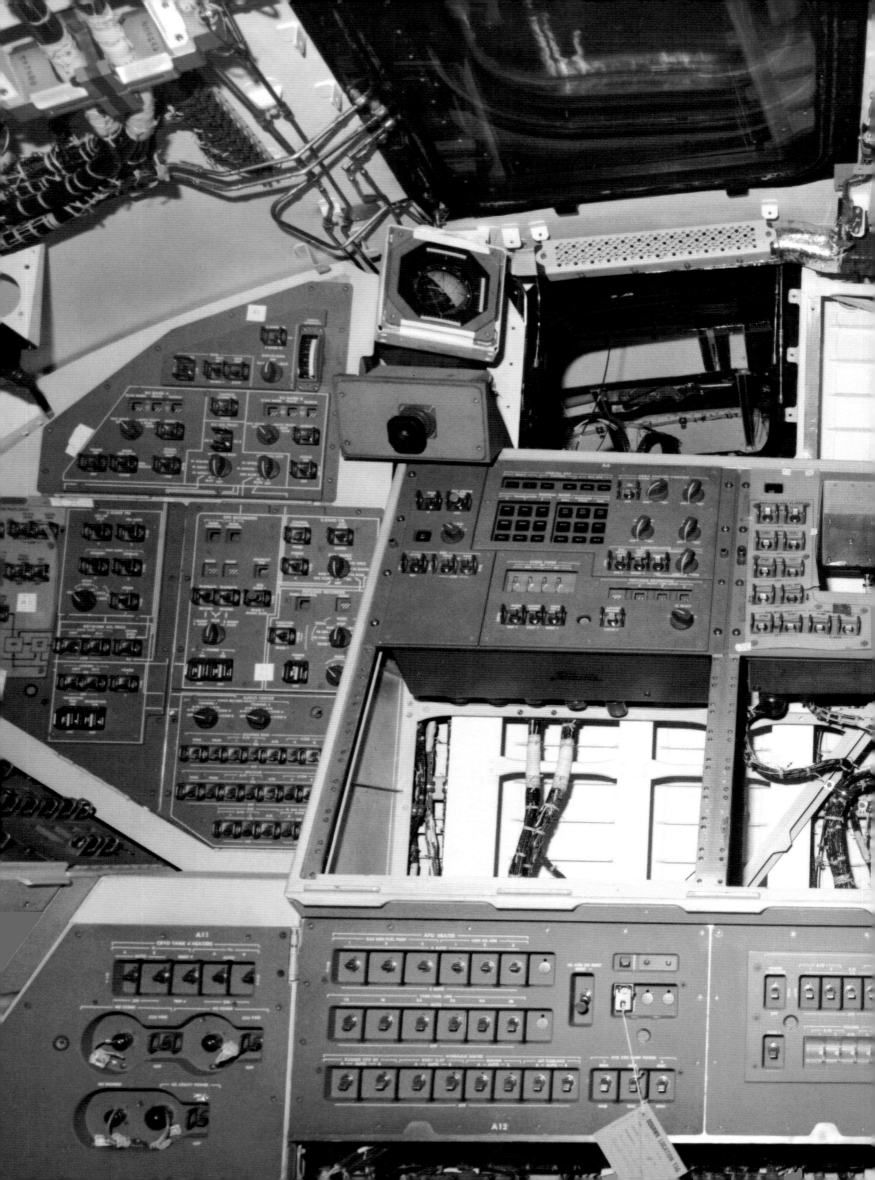

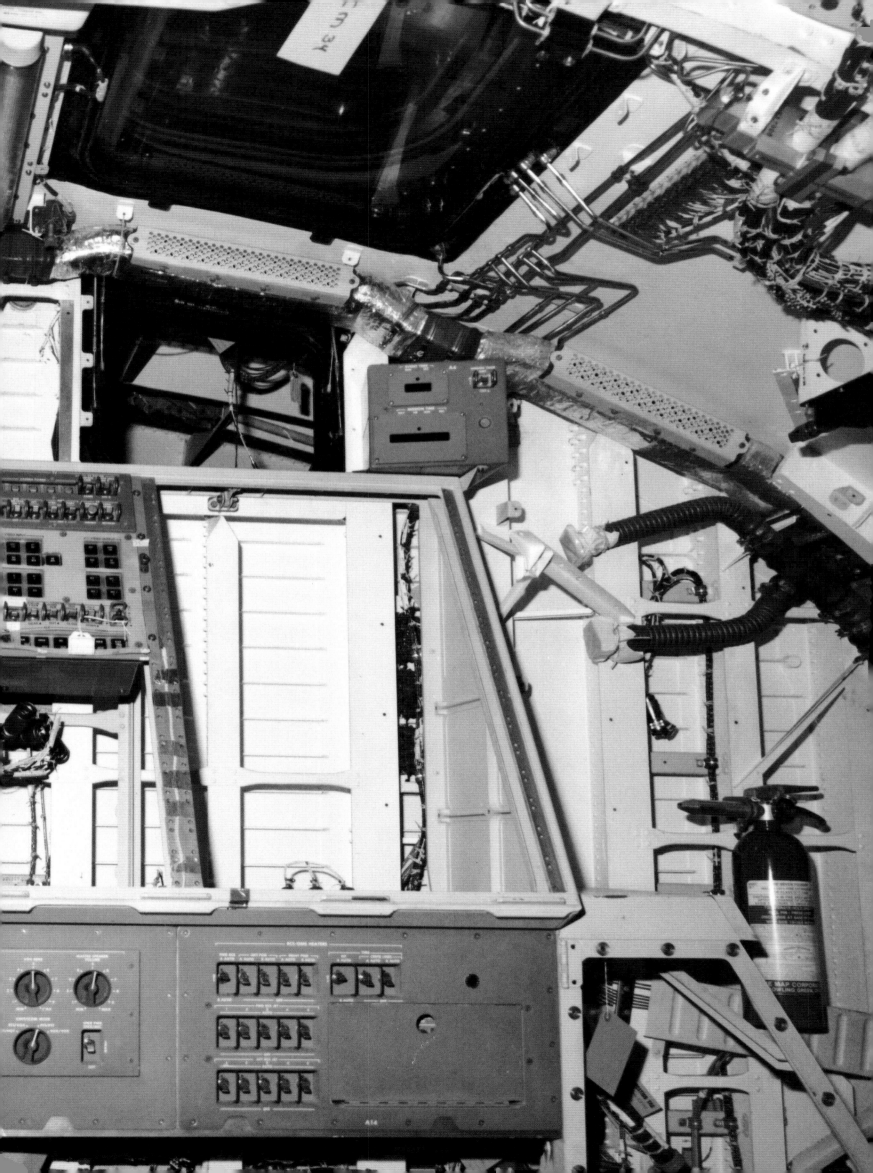

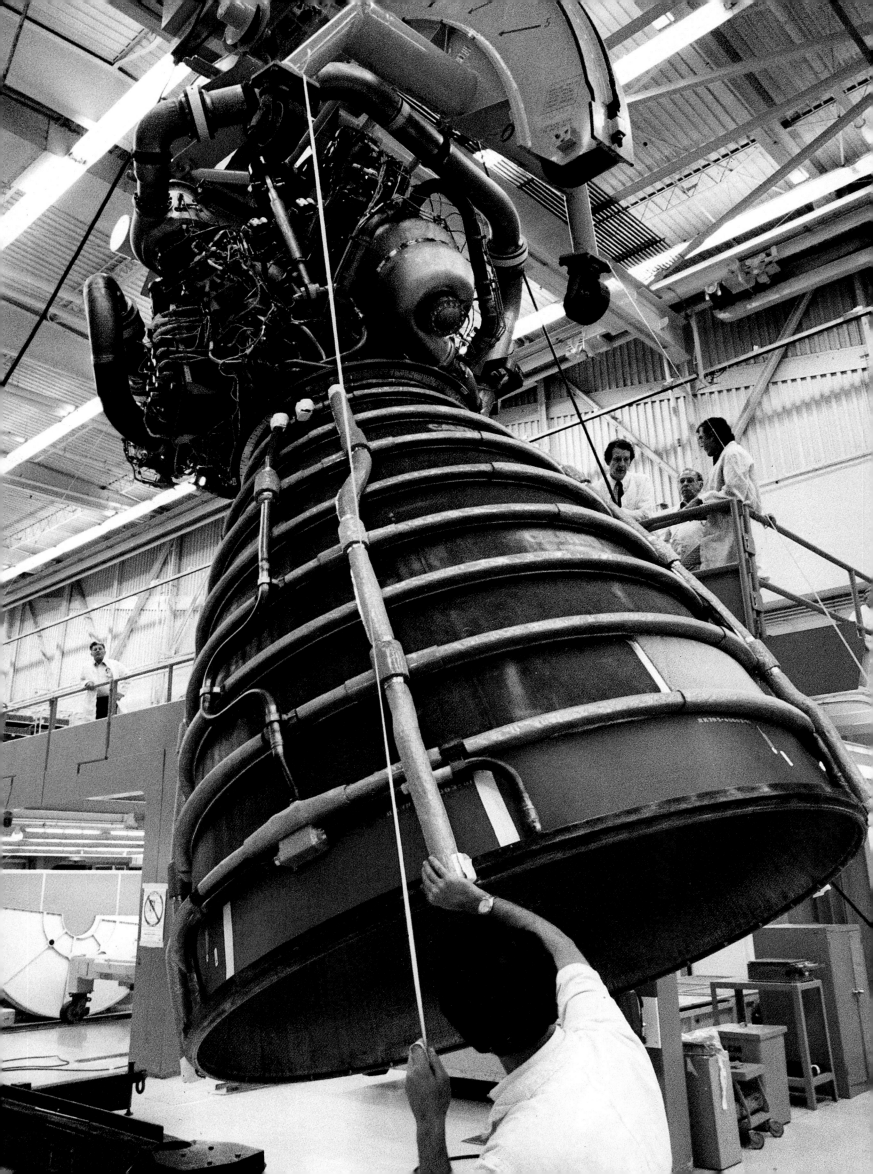

operation of the Orbiter, and *Columbia*'s first four orbital test flights in 1981 and 1982 were flown by a two-man crew. For operational missions the Orbiter is designed to carry up to five additional crew members, who may include NASA mission specialists or payload specialists associated with NASA, the US military, or a representative of an aerospace firm or a foreign country whose payload is being carried. Flights carrying Spacelab, for example, usually carry at least one European Space Agency payload specialist.

The seven-person Orbiter crew is the optimal number for general comfort on long space flights. For this reason, crews of five or six are most common, although *Challenger* carried eight people on a seven-day mission in October 1985. In the case of an emergency, such as the rescue of a stranded Orbiter crew, an even larger number could be carried, but crews in excess of seven persons will be rare.

Aft of the payload bay, below the tail, are the Orbiter's three main engines. Built by the Rocketdyne division of Rockwell International, each is capable of 375,000 pounds of thrust at sea level. These are liquid-fuel rockets capable of 7.5 hours of continuous use without maintenance or overhaul, and supplement the Shuttle's solid rocket boosters.

(continued on page 85)

A high-performance Space Shuttle main engine developed and produced by Rockwell is readied for delivery to NASA's National Space Technology Laboratories in Bay St Louis, Mississippi.

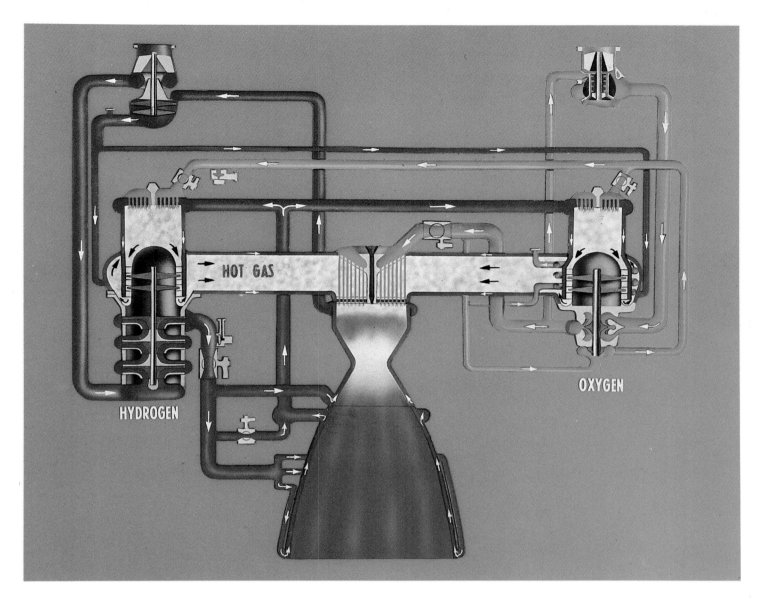

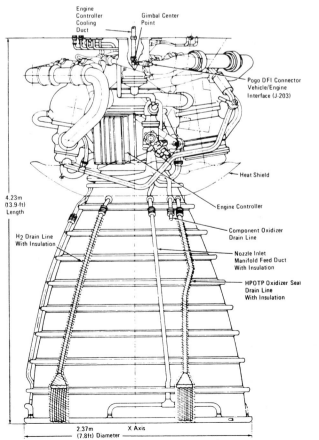

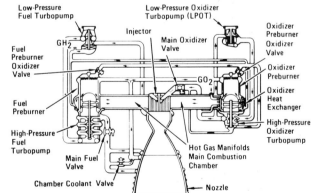

The Space Shuttle main engine under construction at Rockwell in 1982. The Orbiter's three main engines are located at the aft fuselage below the tail in a triangular pattern. These liquid-fuel engines are capable of 7.5 hours of continuous use without maintenance or overhaul. Since they are used on average for only eight minutes per mission, they can be used for over 50 missions unless especially heavy payloads are being lifted into orbit. The diagrams show (*left*) the controller side of the main engine and (*above, both*) the main engine flow. Oxidizer from the external tank enters the orbiter at the LO_2 feedline disconnect valve, then into the Orbiter LO_2 feedline manifold. There it branches into three parallel paths, one to each engine.

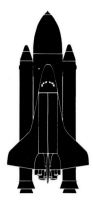

The Mission Profile

Atypical Space Shuttle mission begins with a launch from pad 39A or 39B at the Kennedy Space Center in Cape Canaveral, Florida. It lasts between six and eight days, and ends with a runway landing at either Edwards AFB, California or at Kennedy. After 1986, launches were also possible from Vandenberg AFB, California and landings have been made at other sites, such as White Sands, New Mexico, when weather precluded landing at either Edwards or Kennedy. Operational missions have lasted as few as three and as many as ten days, but the Orbiter is designed to be capable of missions lasting as long as a month.

In a typical mission, the final countdown begins five hours ahead of the launch, with the entire Space Shuttle Transportation System vehicle installed on the launch pad. A half hour later technicians begin filling the external tank. The crew enter the Orbiter an hour and 50 minutes ahead of the launch and begin their final systems check. Four seconds before the solid rocket boosters are ignited, the Orbiter's on-board computer ignites the Orbiter's own liquid-fuel rockets. Seven seconds later, the Space Shuttle lifts off the launch pad. A minute later the Shuttle reaches the speed of sound and is accelerating upward. Two minutes after launch

(continued on page 99)

Columbia, with the unpainted fuel tank, thunders into space on mission STS-3 on 22 March 1982, carrying Gordon Fullerton and Jack Lousma. It was the third flight of the Shuttle program and the third for *Columbia.*

Overleaf: Columbia lifts off Pad 39A on 27 June 1982, bound for a seven-day earth-orbital mission and the final developmental or pre-operational flight for the Space Transportation System. Mission STS-4 was crewed by commander Ken Mattingly and pilot Henry Hartsfield, Jr.

Left: The awesome thrust of the three main engines and the two solid rocket boosters. The main engines each deliver between 375,000 (sea level) and 470,000 (high altitude) pounds of thrust. The solid rocket boosters are solid fuel, pure raw power, and will burn continuously until they burn out.
Above: The Orbiter rolls over in typical head-down position seconds after launch, with its belly pointed away from the earth.

Left: This aerial shot provides a good view of the Kennedy Space Center launch complex as a Space Shuttle claws its way skyward from Pad 39A. *Above:* The Space Shuttle lights up the predawn sky as it lifts away from Pad 39A to begin the STS-8 mission in 1983 with the first nighttime launch of a Shuttle. The mission also ended in darkness, with a nighttime touchdown at Edwards AFB.

Overleaf: Hoot Gibson was the pilot for mission 41-B in 1984, his first. He is shown here helmeted as the Orbiter arrives in earth orbit prior to undertaking the mission. Upon completion of the mission, the Orbiter must re-enter the earth's atmosphere and descend to 45,000 feet before it can be controlled like an airplane. Gibson is seen here reading out the complicated flight data that are then input into the control panels.

91

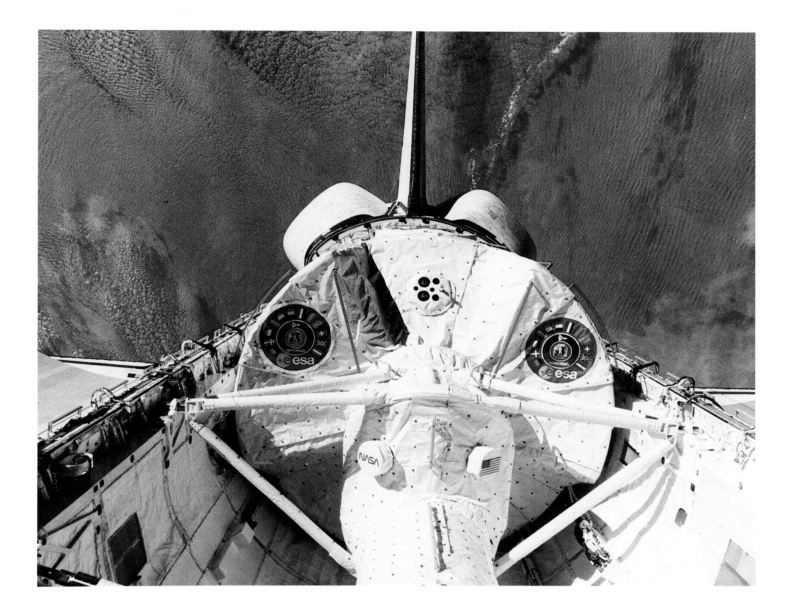

Left: Charles Walker, 41-D payload specialist, closes a stowage area for biological samples supporting the continuous flow electrophoresis systems experiment in the mid deck of *Discovery.* In September 1984 the McDonnell Douglas engineer conducted a televised tour of the refrigerator-sized processing facility for an audience on earth.

Above: The Spacelab module designed by the European Space Agency is marked with all the flags of the member nations. The laboratory, which takes up most of the payload bay, made four flights aboard the Shuttle between November 1983 and November 1985, carrying a variety of experiments.

Left: Wubbo Ockels, a Dutch payload specialist from the European Space Agency, accompanied mission 61-A aboard *Challenger,* launched on 30 October 1985. The ESA Spacelab was being flown for the fourth time, with the payload designation Spacelab D-1. *Above:* An egg floats in the zero-gravity atmosphere of the Space Shuttle during mission 61-A. Henry Hartsfield, Jr is in the background.

the Shuttle reaches an altitude of 150,000 feet. At this point the solid rocket boosters have exhausted their fuel and are jettisoned to be recovered for reuse. Eight minutes into the flight, at an altitude of 380,000 feet, the external fuel tank is jettisoned and the Orbiter maneuvers into orbit. Throughout this period, the Orbiter is traveling downrange away from the launch site as well as upward, so that by the time the fuel tank is jettisoned, it is over the Indian Ocean, about a half orbit into its flight.

Once in space, the Orbiter's flight crew maneuver it from its initial elliptical orbit into a circular orbit. The typical operational altitude for Shuttle missions is 250 miles, roughly 130 miles above the recognized edge of outer space. During the first five years of the Space Shuttle program, the highest altitude achieved was 320 miles, about half the Orbiter's theoretical maximum, reached by *Atlantis* in October 1985.

During the space flight portion of the mission, the Orbiter operates with its payload bay doors open to avoid heat buildup within the bay. The Orbiter usually operates in a head-down position, with the belly of the Orbiter pointed away from the earth.

Re-entry at the end of the mission typically begins over a point 5000 miles from the landing site. The Orbiter re-enters the earth's atmosphere

(continued on page 103)

The Morelos-B satellite, the second in a series of communications satellites for Mexico, was initiated by Sherwood Spring, 61-B mission specialist on 27 November 1985. The payload for 61-B also included the prototype pieces that might later be used for a permanent manned space station. The pieces were called EASE, 'experimental assembly of structures in EVA.'

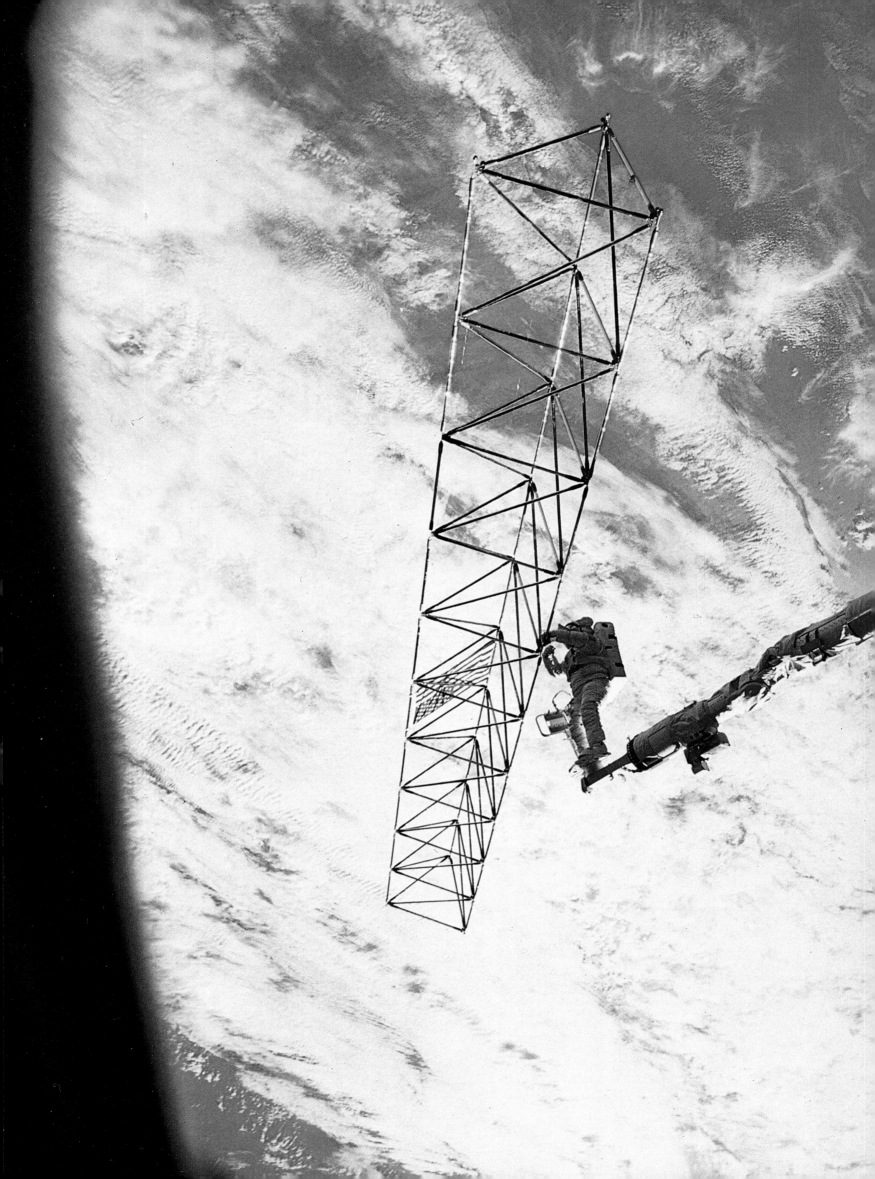

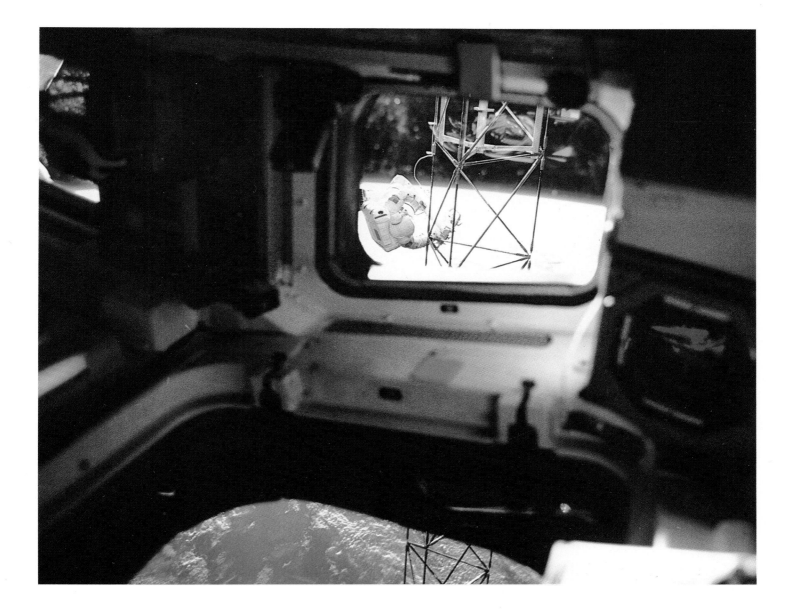

Left: Sherwood Spring, firmly attached to the remote manipulator system by the mobile foot restraint, during 61-B extravehicular activity in November 1985. He was photographed through the window of *Atlantis* (*above*) as he was completing construction of the prototype ACCESS modules, another possible type of structural element for NASA's space station. The idea was that they would try out various components that might later be used in construction of the real space station.

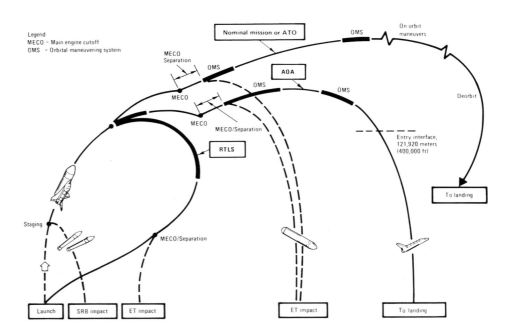

at over 16,000 mph, during which time the atmospheric friction heats the Orbiter's exterior surface to over 2500 degrees Fahrenheit. The ship is maneuvered by small roll and yaw jets until it has descended to an altitude of 45,000 feet and slowed to a speed of about 2500 mph. By this time the flaps and ailerons are functional and the Orbiter can be controlled like an airplane. The landing gear is extended when the craft is about 90 feet above the ground and about a half mile from the end of the runway. With the gear deployed, the Orbiter touches down at a speed of just over 200 mph.

Left: Lining up for a landing on the Kennedy Space Center runway, right back where we started from. Pilots prefer Edwards AFB in the California desert because the runway is longer, but landing here saves turnaround time. The diagram (*above*) shows the Space Shuttle abort and normal mission profile.

Traveling at 200 mph at touchdown, the Orbiter *Challenger* lands at Edwards AFB with a T-38 chase plane flying a parallel course.

The Space Shuttle Flight Log

Test Flights in Earth Atmosphere

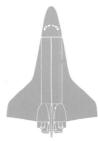

A

Designation: Free-Flight Approach and Landing Test 1
Launch date: 12 August 1977
Orbiter: Enterprise (OV-101)
Crew: Gordon Fullerton and Fred Haise, Jr

B

Designation: Free-Flight Approach and Landing Test 2
Launch date: 13 September 1977
Orbiter: Enterprise (OV-101)
Crew: Joe Engle and Richard Truly

C

Designation: Free-Flight Approach and Landing Test 3
Launch date: 23 September 1977
Orbiter: Enterprise (OV-101)
Crew: Gordon Fullerton and Fred Haise, Jr

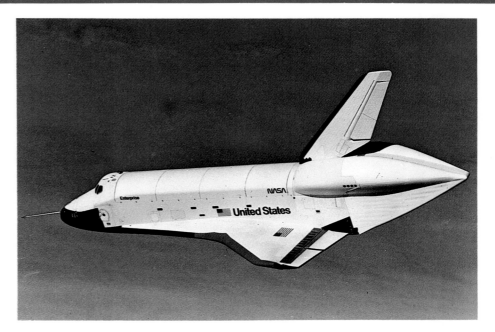

Enterprise, a nonspace-flight test vehicle, made its first flight on 12 August 1977 with Gordon Fullerton and Fred Haise, Jr at the controls on Free-Flight Approach and Landing Test 1. The tail cone covers the main engine nozzles.

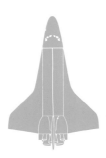

D

Designation: Free-Flight Approach and Landing Test 4
Launch date: 12 October 1977
Orbiter: Enterprise (OV-101)
Crew: Joe Engle and Richard Truly

E

Designation: Free-Flight Approach and Landing Test 5
Launch date: 26 October 1977
Orbiter: Enterprise (OV-101)
Crew: Gordon Fullerton and Fred Haise, Jr

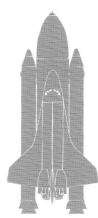

1

Designation: STS-1
Launch date: 12 April 1981
Orbiter: Columbia (OV-102)
Duration in space: Two days
Crew: Robert Crippen and John Young

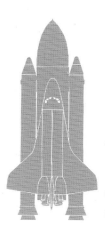

2

Designation: STS-2
Launch date: 12 November 1981
Orbiter: Columbia (OV-102)
Duration in space: Two days
Crew: Joe Engle and Richard Truly

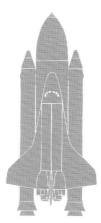

3

Designation: STS-3
Launch date: 22 March 1982
Orbiter: Columbia (OV-102)
Duration in space: Eight days
Crew: Gordon Fullerton and Jack Lousma

The historic first launch of *Columbia* on mission STS-1 took place on 12 April 1981 with Robert Crippen and John Young as crew. This was the first use of solid rockets in manned space flight, and two days later *Columbia* became the first US spacecraft to land on land.

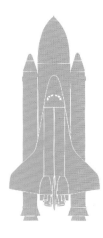

4

Designation: STS-4
Launch date: 27 June 1982
Orbiter: Columbia (OV-102)
Duration in space: Seven days
Crew: Henry Hartsfield, Jr and Thomas Mattingly II

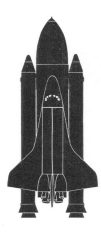

5

Designation: STS-5
Launch date: 11 November 1982
Orbiter: Columbia (OV-102)
Duration in space: Five days
Crew: Joseph Allen, Vance Brand, William Lenoir and
  Robert Overmyer

6

Designation: STS-6
Launch date: 4 April 1983
Orbiter: Challenger (OV-99)
Duration in space: Five days
Crew: Karol Bobko, Story Musgrave, Donald Peterson
  and Paul Weitz

7

Designation: STS-7
Launch date: 18 June 1983
Orbiter: Challenger (OV-99)
Duration in space: Six days
Crew: Robert Crippen, John Fabian, Frederick Hauck,
  Sally Ride and Norman Thagard

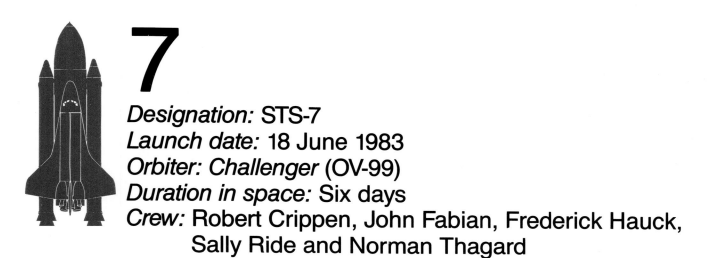

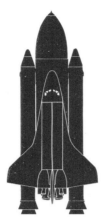

8

Designation: STS-8
Launch date: 30 August 1983
Orbiter: Challenger (OV-99)
Duration in space: Six days
Crew: Guion Bluford, Jr, Daniel Brandenstein, Dale Gardner,
William Thornton and Richard Truly

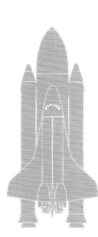

9

Designation: STS-9 (41-A)/Spacelab 1
Launch date: 28 November 1983
Orbiter: Columbia (OV-102)
Duration in space: Ten days
Crew: Owen Garriott, Byron Lichtenberg, Ulf Merbold
(West Germany), Robert Parker, Brewster Shaw, Jr
and John Young

10

Designation: 41-B
Launch date: 3 February 1984
Orbiter: Challenger (OV-99)
Duration in space: Eight days
Crew: Vance Brand, Robert Gibson, Bruce McCandless,
Ronald McNair and Robert Stewart

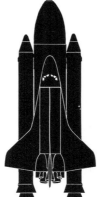

11

Designation: 41-C
Launch date: 6 April 1984
Orbiter: Challenger (OV-99)
Duration in space: Seven days
Crew: Robert Crippen, Terry Hart, George Nelson,
    Francis Scobee and James van Hoften

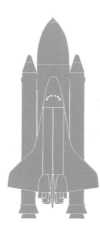

12

Designation: 41-D
Launch date: 30 August 1984
Orbiter: Discovery (OV-103)
Duration in space: Five days
Crew: Michael Coats, Henry Hartsfield, Jr, Steven Hawley,
    Richard Mullane, Judy Resnik and Charles Walker

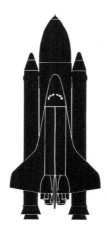

13

Designation: 41-G
Launch date: 5 October 1984
Orbiter: Challenger (OV-99)
Duration in space: Eight days
Crew: Robert Crippen, Marc Garneau (Canada), David
    Leetsma, Jon McBride, Paul Scully-Power,
    Sally Ride and Kathryn Sullivan

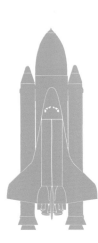

14

Designation: 51-A
Launch date: 8 November 1984
Orbiter: Discovery (OV-103)
Duration in space: Nine days
Crew: Joseph Allen, Anna Fisher, Dale Gardner,
Frederick Hauck and David Walker

15

Designation: 51-C
Launch date: 24 January 1985
Orbiter: Discovery (OV-103)
Duration in space: Three days
Crew: James Buchli, Ellison Onizuka, Gary Payton
and Loren Shriver

16

Designation: 51-D
Launch date: 12 April 1985
Orbiter: Discovery (OV-103)
Duration in space: Five days
Crew: Karol Bobko, E J 'Jake' Garn, David Griggs, Jeffrey
Hoffman, Rhea Seddon, Charles Walker
and Donald Williams

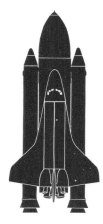

17

Designation: 51-B/Skylab 3
Launch date: 29 April 1985
Orbiter: Challenger (OV-99)
Duration in space: Seven days
Crew: Frederick Gregory, Don Lind, Robert Overmyer,
Norman Thagard, William Thornton, Lodewijk
van den Berg (Netherlands) and Taylor Wang

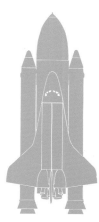

18

Designation: 51-G
Launch date: 17 June 1985
Orbiter: Discovery (OV-103)
Duration in space: Seven days
Crew: Patrick Baudry (France), Daniel Brandenstein, John
Creighton, John Fabian, Shannon Lucid, Steven Nagel
and Salman Abdel-aziz Al-Saud (Saudi Arabia)

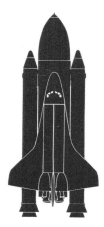

19

Designation: 51-F/Spacelab 2
Launch date: 29 July 1985
Orbiter: Challenger (OV-99)
Duration in space: Nine days
Crew: Loren Acton, John Bartol, Roy Bridges, Jr,
Anthony England, Gordon Fullerton, Karl Henize
and Story Musgrave

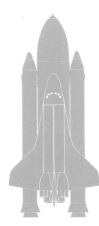

20

Designation: 51-I
Launch date: 27 August 1985
Orbiter: Discovery (OV-103)
Duration in space: Seven days
Crew: Richard Covey, Joe Engle, William Fisher,
    Michael Lounge and James van Hoften

21

Designation: 51-J
Launch date: 3 October 1985
Orbiter: Atlantis (OV-104)
Duration in space: Four days
Crew: Karol Bobko, Ronald Grabe, David Hilmers,
    William Pailes and Robert Stewart

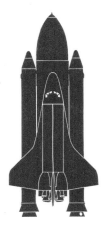

22

Designation: 61-A/Spacelab D-1
Launch date: 30 October 1985
Orbiter: Challenger (OV-99)
Duration in space: Seven days
Crew: Guion Bluford, Jr, James Buchli, Bonnie Dunbar,
    Reinhard Furrer (West Germany), Henry Hartsfield, Jr,
    Ernst Messerschmid (West Germany), Steven Hagel
    and Wubbo Ockels (Netherlands)

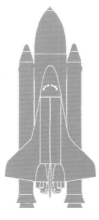

23

Designation: 61-B
Launch date: 26 November 1985
Orbiter: Atlantis (OV-104)
Duration in space: Seven days
Crew: Mary Cleave, Rodolfo Neri (Mexico), Bryan O'Connor, Jerry Ross, Brewster Shaw, Jr, Sherwood Spring and Charles Walker

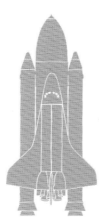

24

Designation: 61-C
Launch date: 12 January 1986
Orbiter: Columbia (OV-102)
Duration in space: Six days
Crew: Charles Bolden, Jr, Robert Cenkar, Franklin Chang-Diaz, Robert Gibson, Steven Hawley, Bill Nelson and George Nelson

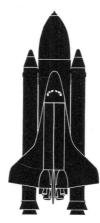

25

Designation: 51-L
Launch date: 28 January 1986
Orbiter: Challenger (OV-99)
Duration in space: None
(vehicle exploded 73 seconds after launch)
Crew: Gregory Jarvis, Christa McAuliffe, Ronald McNair, Ellison Onizuka, Judy Resnik, Francis Scobee and Mike Smith

The Orbiters

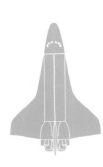

Enterprise (OV-101)
First flown in the earth's atmosphere on 12 August 1977; last flown on 26 October 1977. Never intended for space flight. Retired in 1985.

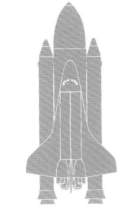

Columbia (OV-102)
First flown in space on 12 April 1981. Out of service between November 1983 and January 1986.

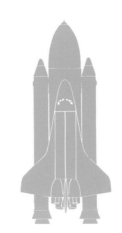

Challenger (OV-99)
Not originally intended for space flight, but later converted. First flown in space on 4 April 1983, destroyed on 28 January 1986, 73 seconds into its tenth flight.

Discovery (OV-103)
First flown in space on 30 August 1984.

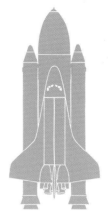

Atlantis (OV-104)
The last of the originally planned group of four spaceworthy Orbiters. First flown on 3 October 1985.

Notes on Space Shuttle Transportation System (STS) Mission Designations

The first five missions were Free-Flight Approach and Landing Tests flown within the earth's atmosphere by *Enterprise* in 1977. The first nine space flights of the system, between 1981 and 1983, were divided between *Columbia* and *Challenger* and were designated STS-1 through 9 consecutively. Mission 9 also carried the second designation, 41-A, which was the first application of the new nomenclature system under which all subsequent Shuttle missions were designated.

In the new system, the first digit stands for the NASA *fiscal* year in which the mission was *originally scheduled* (NASA fiscal years begin in October). The second digit refers to the mission launch site, with 1 assigned to Kennedy Space Center and 2 assigned to Vandenberg AFB. The third digit is a letter indicating the order in which the mission was *originally* scheduled within the fiscal year. Thus 41-A is read as the first mission scheduled to be launched in fiscal year 1984 from Kennedy Space Center.

Under this system a mission can be designated before its launch and can retain the same designation throughout its planning stages, even if it falls behind schedule. For example, mission 51-B was planned to follow 51-A, but actually took place after 51-C and 51-D. Missions can also be canceled without affecting the designations of other missions. For example, mission 41-G followed mission 41-C when missions 41-E and 41-F were deleted from the roster.

Dedication

The publishers wish to dedicate this work to the seven brave crewmembers of Space Shuttle mission 51-L, lost aboard the Orbiter *Challenger* on 28 January 1986: Gregory Jarvis, Christa McAuliffe, Ronald McNair, Ellison Onizuka, Judy Resnik, Francis Scobee and Mike Smith.

Acknowledgements

The Author wishes to thank Mike Gentry at NASA's Johnson Space Center and Sue Cometa at Rockwell International for their assistance in making this book possible.

Picture Credits

All photographs were supplied by **Rockwell International, Space Transportation Systems Division,** except for the following:

NASA: Front Cover, 8, 10, 11, 30-31, 33, 34-35, 36-37, 38, 40-41, 44, 46-47, 48, 50-51, 53 both, 65, 67 bottom, 68-69, 70, 71 both, 73 both, 76, 77, 83 bottom both, 91, 94, 96, 97, 98, 100, 101, 103

Ralph Morse, Time-Life: 52

Edited by Susan Garratt
Designed by Bill Yenne

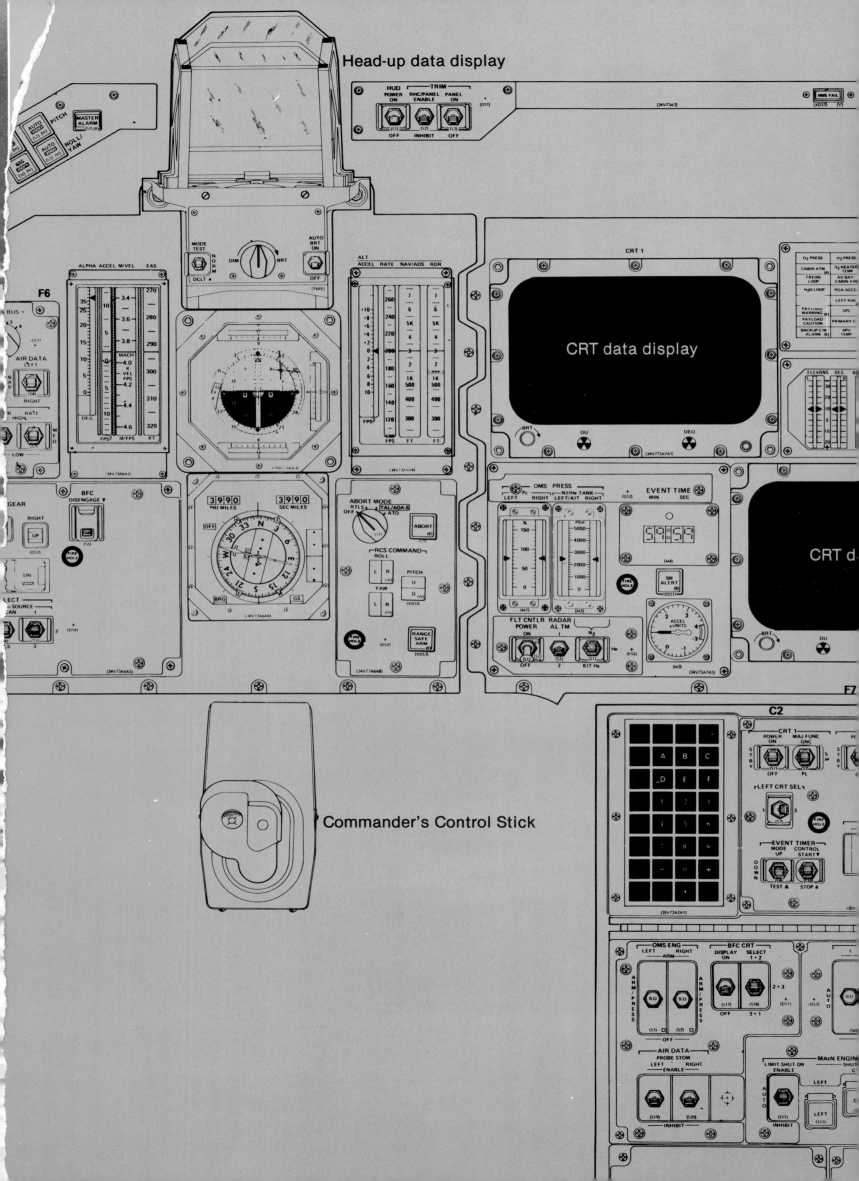

Head-up data display

Commander's Control Stick

CRT data display

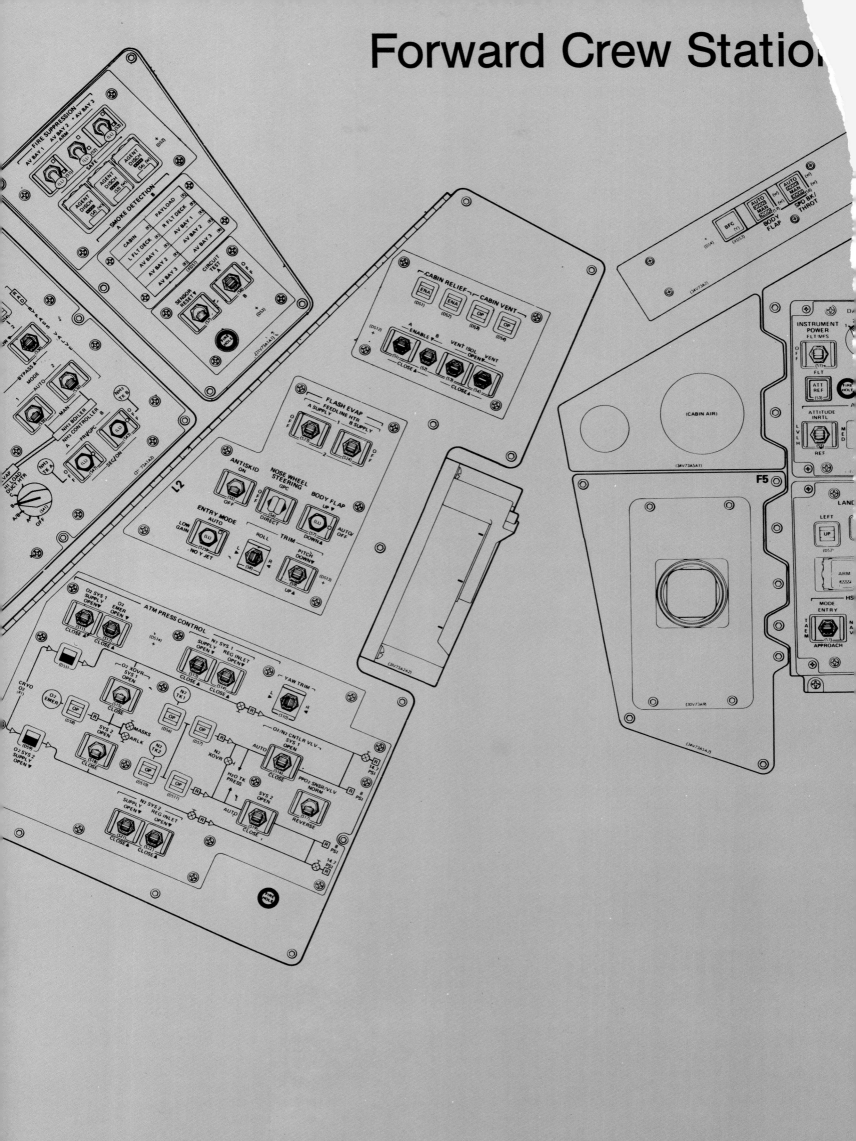

Aft Crew Station (port side)

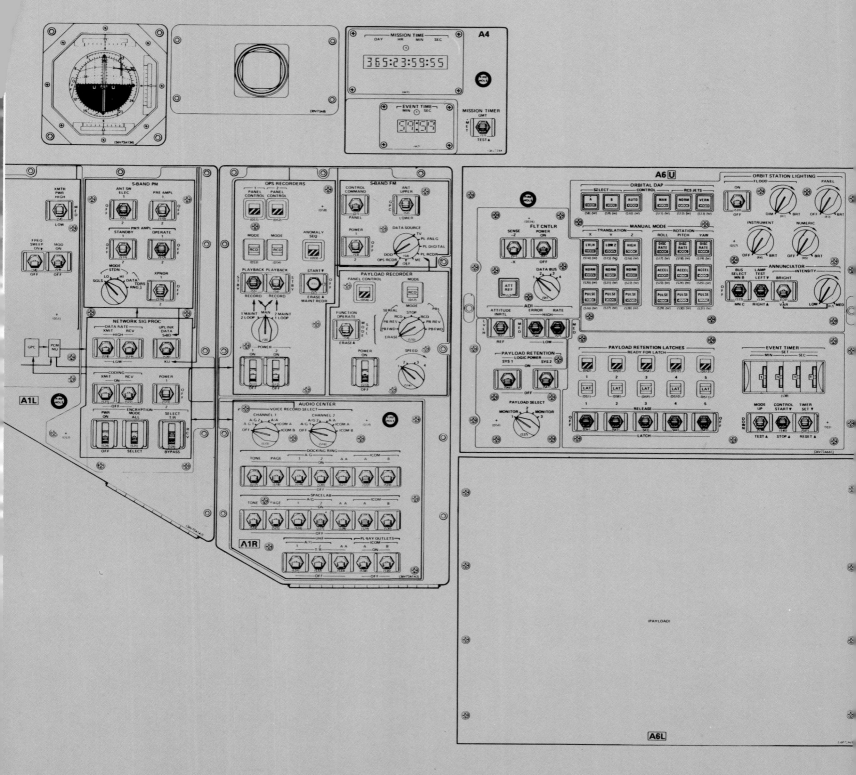

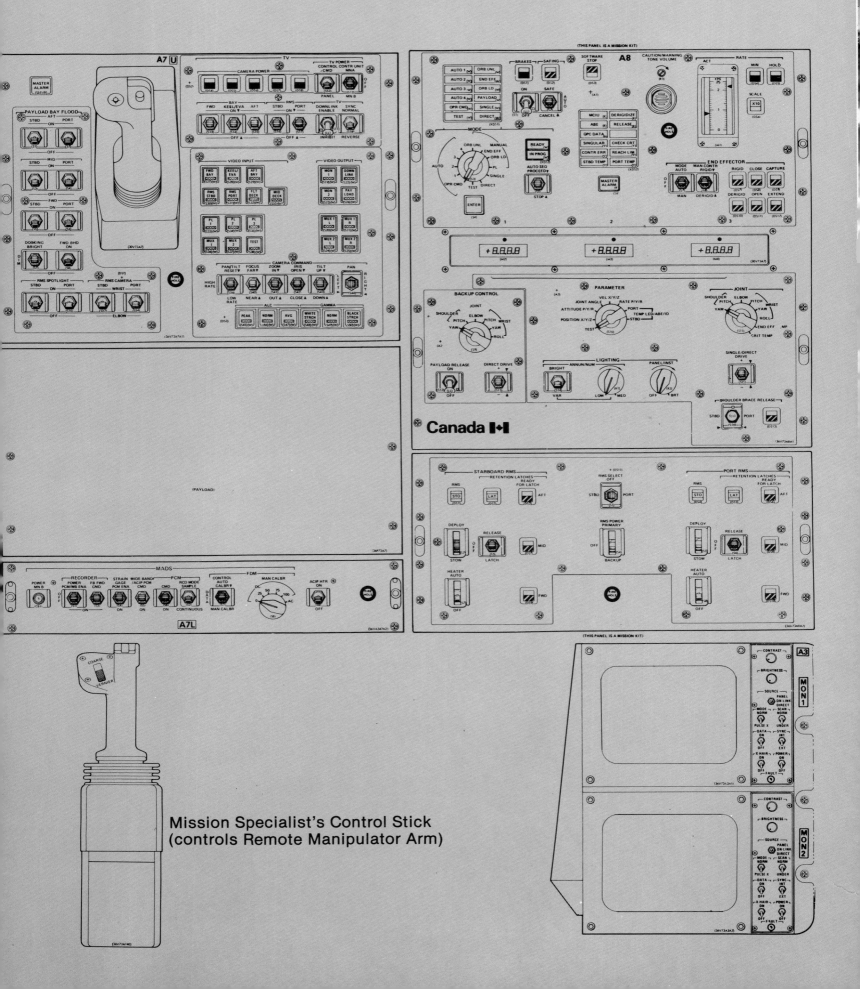

Mission Specialist's Control Stick (controls Remote Manipulator Arm)

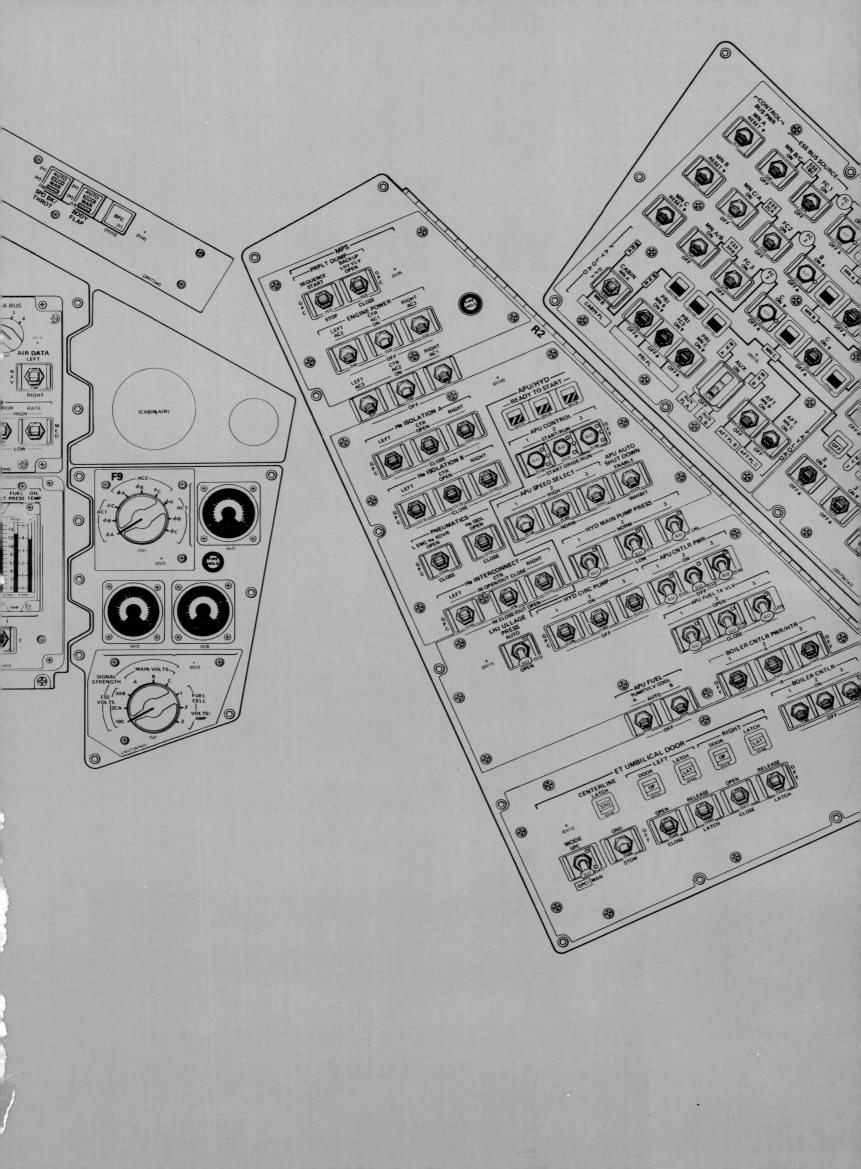

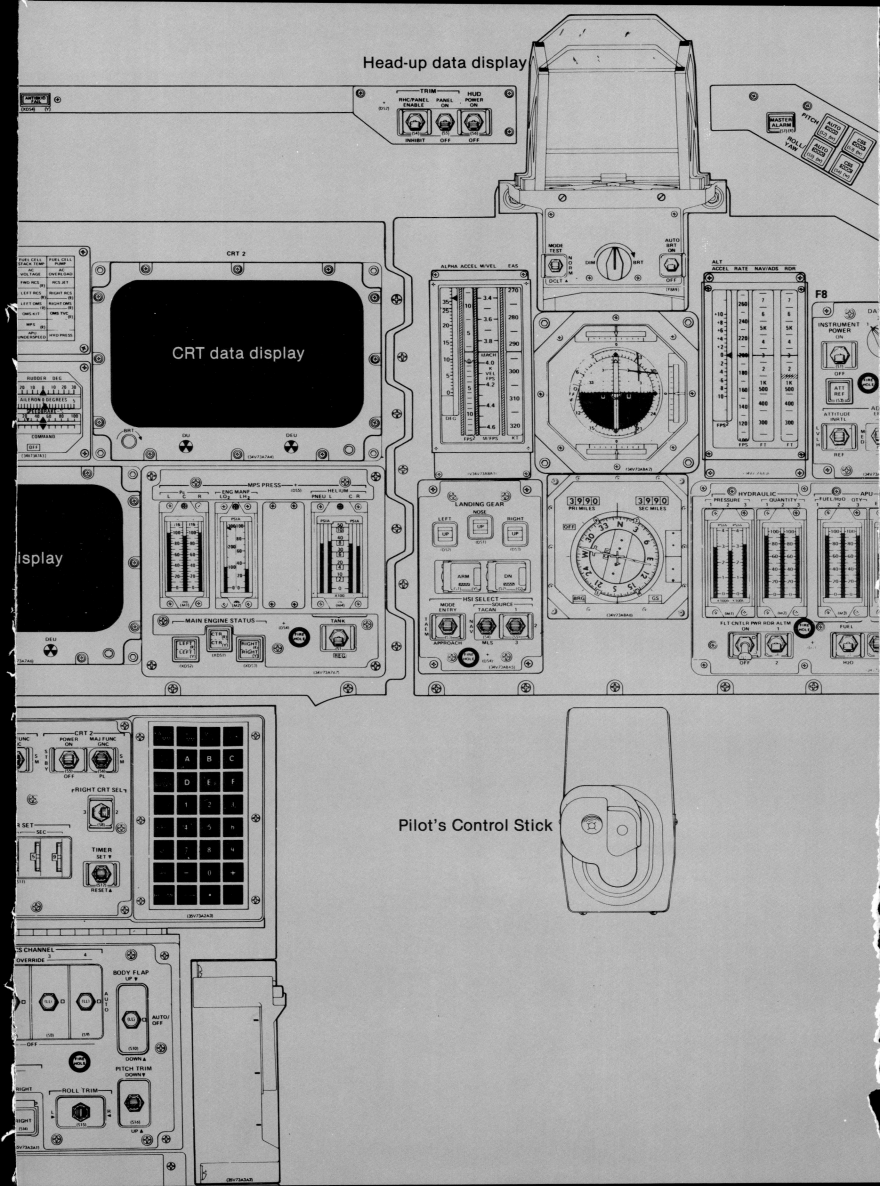

Head-up data display

CRT data display

Pilot's Control Stick

F8